Praise for *An Introver*

"*An Introvert Learns to Fly* shows the fullness of Peterson's life as a son, husband, father and innovative business leader."

—Fred Faxvog, friend and Honeywell colleague

"Peterson ran our research labs the way he played golf. Deep in thick woods and weeds, he always shot for the green. He told me, 'Nothing is impossible, but miracles cost a million dollars.'"

—Jim Lenz, Honeywell colleague, sometimes professor,
John Deere executive, and friend

"A wise scientist! A man of character who inspired confidence. He did things worth doing, and still does. A great boss!"

—Carol Warne, longtime Executive Assistant

"Peterson describes how a shy quiet kid from Minnesota can survive and become aware and assertive at a top university like Caltech, surrounded by brilliant East- and West-coast types."

—Erik Lippa, MD PhD, Caltech roommate

"*An Introvert Learns to Fly* provides an inside look at Ron's career and how things worked in the heyday of Honeywell in Minnesota."

—Deb Holmgren, first cousin

"This book will help you explore and relish life, as I discovered with Ron when we were kids." —Marsha Holmgren, first cousin

"Peterson's parents and teachers urged him to explore his imagination and love of science. With encouragement, a shy child can become part of history."

—Phyllis Moore, Author and friend

An Introvert
Learns to Fly

An Introvert Learns to Fly

A Memoir of Timidity, Panic, Science, Leadership, and Love

Ronald E. Peterson

BOOKS

An Introvert Learns to Fly
Copyright © 2018 by Ronald E. Peterson

ISBN 978-0-9997035-0-2

Order online at itascabooks.com

Inquiries may be sent to:
PTB Books
535 Tomlyn Avenue
Shoreview, MN 55126
ronsreadingroom.com

Edited by Pat Morris

For Miriam, Amy, and Kels,
and all the others who helped.

Table of Contents

PART III: GIVING YEARS

An Introvert
Learns to Fly

PART I

JUMPING YEARS

These are the stories I tell, which seem to pop up again and again. A few details may be tweaked by my mischievous memory, but they're told in a way that people seem to enjoy.

My name is Ron Peterson, and in most ways I'm a normal person. For some reason I've led an interesting and possibly consequential life. From early on I was a serious introvert, shy and cautious with people. Also hidden within was a spark, sometimes a fire, which urged me to do more than just get along, to fly, to change the world if possible. These competing forces shaped my life.

My years have been divided into three parts. The first 27 were dedicated to learning, from infancy through my physics PhD. Those years were when I learned to hop around at home and then leave the nest. I call Part I my "Jumping Years."

The middle 27 years were spent as a physicist, top executive, husband, and father. Perhaps they should be called "Real Life," but I'm calling Part II simply the "Working Years." During those years I learned to lead, to fly, and even soar.

I'm retired now, in the last years of my life, doing what I like and passing on as much as I can to my world, neighborhood, children, and grandchildren—teaching others how to fly. These I call Part III, the "Giving Years."

If I die during my 81st year of life, my time on earth will be divided into three equal 27-year portions—very neat and tidy. There is still time to write books. I'm 72, but I have a lot of other projects I'd like to finish as well.

You are free to draw conclusions and morals from these stories of my life.

Being an introvert has aspects that I managed to both exploit and over-come. Learning from my ideas and trials may be of value, especially if you are also an introvert. But I will resist preaching to any of you—I'm not that self-assured.

Our Families

This chapter summarizes the family history of my father Eddie, mother Fern, and some of those connected. I had one sibling, my sister Janis, who was about two years younger than me. My grandparents' families came from Norway, Germany, Denmark, and Sweden.

I don't know much about my father's Swedish mother (Anna née Mattson) except that she died early, at 60 years old or so. I think she was the source of bad heart genes that killed off most of my father's nine siblings.

My dad's dad, Barney, a blacksmith, lived to be 97. All my life I've told a story about Barney's last name, Grenlin, which he later changed to Peterson. Barney emigrated from Norway when his mother remarried. Since he was no longer the eldest son in the new family, he lost the future ownership of the family farm and chose to move to America. When here, after his sec-ond child was born, he changed his name to Peterson, abandoning his step-father's name of Grenlin. Hence, I am a Peterson and literally hundreds of Barney's descendants are Petersons.

We didn't spend much time with my father's relatives. We saw most of those cousins only on a few visits each year to Uncle Art's farm near Gatzke, Minnesota, or Uncle Walter's house in Holt, Minnesota, (population 98). Some summers we would vacation at Art's and watch the cows and help with the harvest. At Walter's we learned about outhouses and water pumps in the kitchen, marveled at his large garden, caught fireflies at night, and walked to town for ice cream at the town's only store. Walter, my father's oldest brother, had acted almost like a father for him. Our cousins Up North were generally much older than us.

Two of my father's siblings, Alma and Ernie, had moved to the Twin Cities. Alma, a compulsive perfectionist, owned two restaurants and became quite wealthy, living on a large lot along the Mississippi. My sister Janis and

I once got in a canoe there and drifted toward a river dam with everyone screaming for help. Paddling furiously, we survived.

Uncle Ernie's family and ours were very close since we shared a duplex with them. Cousins Jerry, Jim, Gary, and Debbie were all roughly our age, and we got to know them well. We lived in the upstairs part of the duplex at 950 Burr Street until I was 16, where I shared a bedroom with Janis.

My mother's parents were Hans and Ida Hansen, who were wonderful people. Our family spent most Sunday afternoons with them. Hans, a very playful man, would chase Janis and me around the house snapping his false teeth in his hands. He was a sheet metal worker with an iron grip. He liked to shake hands and drive you to your knees. Ida was a classic grandmother, exceedingly warm and nurturing. She taught us all how to knit, crochet, and cross stitch, at which I became fairly skilled.

Their large dining room brings back deep memories. There was a shelf a couple feet below the ceiling on two walls that was covered with family photos and grandchildren's handiwork, including my artwork. A large cabinet held hundreds of salt and pepper shakers from their trips. A soft couch was located by the windows and between an old-fashioned (foot-pumped) sewing machine on one side and a canary cage on the other.

Sunday afternoon dinners in that room often included my aunts, Ruby and Myrtle, or uncle Del, with cousins Marsha, Deb, Dennis, Steve, Jim, or Judy scattered about. Hot dish, roast beef, or ham was often the main course. The men would play Canasta after eating, with a 25-cent bet on each corner of the card table. It was a great honor when I was finally allowed to play with the adults.

When very young, we grandkids mostly dug in the backyard dirt or snow, or ran around the house or down to the corner on the sidewalk. By school age, we'd play "post office" upstairs, each going to a separate room, slipping notes under the door, and awaiting a reply after a designated postman delivered the note. Young Deb didn't like this game because a stuffed bird mounted on the hallway wall frightened her. When we were older, we developed a family game we call "duple-duple." This multiplayer card game takes solitaire to its limits, with everyone playing on the common building stacks in the middle. Up to ten people have played this game at once on the floor, aggressively slapping cards down before anyone else is able. There are actually rules to prevent injury.

My very first recollection from Hans and Ida's house is of a small windup train going around in a circle on the dining room table at Christmas. I was perhaps about 18 months old at the time. A year later, I started regularly climbing the pillars that separated the living room from the dining room in their house so I could touch the ceiling. Possibly because of that, Hans often gave me a lump of coal for my birthday. He also sometimes gave me a potato (along with nicer gifts) for Christmas and told me potatoes would grow from my ears if I didn't keep them clean.

Early Jumps

I call the first 27 years of my life the jumping period because I took many leaps to discover the unknown and to learn quickly and deeply. Sometimes I succeeded, sometimes I failed, and sometimes I got hurt, but taking risks is still the best way for me.

And sometimes, I got my name in the newspaper. My first act of public notoriety occurred at age three. My mother, Fern, knew that I liked to explore, so she told me that I shouldn't leave the yard without someone older. Well, one of the neighborhood girls was four years old so I asked her and her sister to join me on a walk. From our house on the near East Side, the plan was to walk to Hans and Ida's on St. Paul's West Side, about 5.5 miles, the route our family had traveled a hundred times. I recall that the elm leaves had begun to fall, and we kicked them as we walked along our Burr Street sidewalk. We crossed the railroad track bridge, cut into Italian Town, past a school, through the warehouse district, and into downtown St. Paul along 7th Street. Turning left onto Robert Street, we managed to reach the bridge over the Mississippi, about three miles total, before we were "arrested."

The police took us in a squad car to their downtown headquarters. We were offered candy, but I turned it down saying, "I don't like that kind." I knew our family's address, and our parents had probably called, so we did eventually get home. Before leaving, we were propped up on the big police desk, a classic design often seen in old movies, with two light balls atop pedestals at either end. This image appeared in the St. Paul paper the next morning, with the two girls smiling and me frowning at the camera. My mother kept this photo in the living room end table until she died, but it was subsequently lost.

When we returned home in a police car, dozens of people ran out to greet us. The entire neighborhood had been searching for us. My mother grabbed me—happy, angry, and crying.

I don't recall a single instance, as I grew, where my mother was verbally angry with me or demeaned me in any way. She mostly just gave me her "disappointed in you" look. And that's after doing some not-so-nice things to Janis—teasing, etc. Mother never punished me for mistakes, although my dad would occasionally snap one or two fingers across the back of my head to get my attention. I don't remember how he felt about the whole "walking to Grandma's" episode.

Grade School

My mother was intelligent, graduating near the top of her high school class, but spent most of her life as a switchboard operator, clerk, or secretary. For several years during the Depression, she was the only family member with a job. She would make the long walk from the West Side bluff and across the Wabasha Bridge each day to downtown St. Paul. Pictures of the family from that time showed everyone looking very thin, almost emaciated. It was at one of my mother's downtown jobs, at G. Sommers Company, working as a phone switchboard operator that she met my father Ed, who enlisted in the Air Force before the start of WWII. Hans and Aunt Ruby had jobs at the Twin Cities ammunition plant when the war started.

My father, Ed, also had a tough life, partly due to speech and grammar issues. His father spoke mostly Norwegian, and Ed (a.k.a. Eddie) had a "Fargo" accent. During the war, he served in the Air Force as a Chemical Warfare Trainer. He would send new recruits into a building filled with tear gas after showing them how to wear gas masks. Inevitably, he said, at least one soldier would come out crying because he had wanted to sample the gas. By the war's end, he was a Technical Sergeant stationed near Salt Lake City, where I was born.

Education was a top priority in our family for both my parents. Janis and

I knew how to count and could even read a little before kindergarten. Our mom stayed at home during those years and gave us our first lessons. Our grade school, Ericsson, was only four blocks from our house at 950 Burr Street in St. Paul, so we walked to school and were able to go home for lunch each day.

I was very lucky and privileged growing up, both by having deeply loving parents and several teachers who were more than just teachers. These key people took the time and interest to care about their students and, for me, acted as personal mentors. Miss Kaye, Mrs. Graves (née Nelson), Miss Andrewson, and Miss Olson all left their mark on me during grade school.

In first grade, I copied an incorrect answer from a classmate. Later, complaining to Miss Nelson that I knew the right answer and should get credit for it, she said something about always trusting myself. She also said that cheating was a very bad idea. My parents must have reinforced this, since I have never forgotten the incident. Surprisingly, Mrs. Graves wrote a congratulatory letter to me when I graduated from high school; some teachers really cared.

When my parents had friends over, I often had to put on a show with my math skills. Dad would put down six numbers, all in the billions, and ask me to add them up. Later, I had to demonstrate multiplying giant numbers. Numbers and math were important, I learned.

In fifth grade, Mrs. Olson invited students and parents to spend the cold fall evenings looking through her telescope and then having cocoa at her house. My parents came along, and the experience created a lifelong interest in astronomy for me. I built my first telescope from parts purchased from Edmund Scientific in seventh grade, around 1957 after the Soviet Sputnik satellite went up. We would look at planets out the 12 top windows of our house's west-facing porch. When the Echo Balloon, the first passive communications satellite, was successfully launched in 1960, I took the scope out to the busy sidewalk at Burr and Case. A dozen people huddled around my telescope as we tried to catch a view of Echo speeding by (a very difficult thing to do with a small telescope), but we saw the balloon clearly in binoculars.

A book report in seventh grade reinforced the "science thing" for me. I'd read a book about electronics, electrons, and Ohm's law, etc., that also described vacuum tubes and how they worked. Most kids stood at the front of the class and maybe had a few photos as part of their book reports—I built a TV. Not a real TV, but a cardboard box with a light bulb in the back. Two paper towel cardboard centers were in the front with a wax paper roll stretched between them. While describing what I had read, I moved the drawings on the wax paper from one image to the next. TV commercials were interspersed with the book report. Mr. Otterness, my teacher, was so surprised he called the principal into the class and I had to present the talk again. He said it was "college material" and wrote a letter to my parents.

Seventh grade was also when I started being selected for "responsible" positions. As captain of the school police patrol, I had to make sure the younger children crossed all the major streets safely on their way to school, and to make sure all the busy corners were covered with other officers, my first supervisory position.

I wasn't always a good kid. On one summer day I shot dried peas out of a straw at passing cars from the front porch. It was great fun until one car screeched to a stop and the driver came running back to our house, pounding at the door. I had hit his mother, who was sitting in the right front seat with the window open. When he started pounding, I hid under my bed and waited it out. I remember my mom calling out for me after he left and I probably got an icy stare and words later.

Summers were spent playing with my best friends, Kenny, Harwood, and Harold, who all lived within two blocks. We played pool and cards on a tiny porch table and organized baseball on the Wilder Playground team. Second base was my position because I couldn't throw well. I was very small in seventh grade, which is odd because in kindergarten I had been one of the tallest. I wouldn't be tall again until the end of high school.

Playing sports and other games with bigger and older children taught me how to compete. One time in Kenny's backyard I wrenched my knee when an older brother fell on it during a football game. Afterward, my mother said I couldn't play football when I got older, which was quite a loss in my mind.

Something very important happened while I sat on the bench during one baseball game at age 11. For the first time I realized that I was in charge of my life. I could be what I wanted—not what my parents or anyone else expected. I didn't know what I wanted at that point, but I would be the person to decide.

The Dump

Mom occasionally referred to it as the Woods, to give it a bit of class, but Kenny, Harwood, the cousins, my sister, and most everyone else called it the Dump. To adult eyes it wasn't much, just a two-block by one-block undeveloped rectangle with houses around the edge. The land was low, not suited for construction, but it had a 20-foot rise across the middle where a dirt road led over to a scary two-story house. Old people used to yell at us from there if we got too close.

To really young eyes, five, six, or seven years old, the Dump was adventure. Our moms finally let us go in there by ourselves. Spring rains filled it up, so we could float on rafts left by bigger kids or climb out over water on fallen elms. Purple-pink weeds soon bloomed, and we caught bees in Skippy peanut butter jars with holes punched in the lid. We'd see how many we could get in one jar, and later, how long it took them to die.

As I got older I learned about many interesting spots: the Italian neighborhood across the tracks, spooky, burned-out Swede Hollow, Centerville Lake where we would bike to go fishing, and the ivy-covered wall behind the Capitol Theater that we'd climb up on Saturday afternoons. But the Dump was a special place where stories oozed out.

Kids shared tales there about the 30 houses around it: about the wheelchair man on the corner, broken collar bones, school projects, girlfriends, sports heroes, and parents who were alcoholics, juicy stories that I've mostly forgotten.

In winter, the Dump was the place to slide after school. The rise in the middle became icy when the snow packed down. We'd run off the road, jumping on our sleds while still in the air, and plummet down the icy hill, or even better, stand on the back boards of our American Flyers holding the cord like the reins on a horse. Getting to the bottom without crashing gave you bragging rights.

After supper we'd cut through the Dump quickly to get to Wilder Playground and its huge ice sheet and blazing lights. We lived for the ice, hockey of course, but also a tag game where leaping into snow piles over your head made you safe. If our feet froze, we'd play basketball. Running home late along the edge of the Dump in the dark, rats would sometimes scurry out.

The Dump had danger, like the deep caves that teenagers dug and covered with boards, the vine-swings over drop-offs, and the places where people threw their junk. We played Davy Crockett, bank robbers, bogeymen, bogeywomen, and sang pirate songs. It was a beautiful place, our Dump.

After 50 years I can still visualize the Dump. I remember where every tree and cave was located. Four more generations of kids have grown up in the old neighborhood and the Dump is still the only green area within miles in all directions. I hope those children have been as blessed. I'm only sure that for us kids of the '50s our East Side Dump was a place where imagination and reality blended perfectly.

Junior High (Grades 8 and 9)

Going to Cleveland Junior High for the first time was a scary event. It involved walking about 10 blocks from home across busy Payne Avenue. Worse, all the children from the other elementary schools would be there, including Farnsworth, the rich kids' school.

In fact, none of the kids at Cleveland were really from rich families; some were middle class at best. But the Payne/Case area where I grew up was clearly lower class. It was, and always has been, a transitional neighborhood, where new immigrants first settled and then left as soon as possible. Swedes, Poles, Italians, Mexicans, and now Hmong have all called the area home.

So all Ericsson kids had reasons to be nervous at our new school. My first algebra test in Miss Opitz's class surprised me. I had studied very hard for that test but was shocked to find I got the highest score, even with "rich kids" now in the same class. That's when it became clear that math might really be my thing.

I signed up for basketball and was later named captain of the team. It's actually very cold playing basketball in those skimpy outfits during a Minnesota winter, and I shivered a lot. During our first game, I scored the first points of the season by making an underhand free throw. I remember people cheering along the sidelines. But I never got over the shakes and, unfortunately, I never scored again for the rest of that season.

How the Jumper Got His Name

During my basketball "career" as captain, I often had to go to center court and listen to the referees explain the rules. It also involved introducing myself to the opposing team captain(s). I remember wondering, sitting on the bench at an away game at Como Junior High, if I should go with Ronald, which seemed too formal, or with Ronnie, as my family and most others called me. Ronnie seemed too cute for junior high. I had a breakthrough, realizing that I could be just "Ron." It was a momentous decision; I have been Ron ever since except when a distant cousin visits, and then that person might call me Ronnie despite my best efforts to correct them.

It is important for you to know why I call this portion of the book the "Jumping Years." When I was quite young I would run down the concrete sides to our front steps and leap as far as possible, measuring the distance with a ruler and recording the results. I had a recurring dream at the time that if I jumped just a little farther each day, eventually I would be able to fly. I hadn't yet learned anything about human physiology, gravity, calculus, or "reductio ad absurdum" logic, for that matter. I imagined how it would feel to be a bird. For a good portion of my early years, I dreamt of flying.

Telescope Number Two

I had an excellent science teacher, Mr. Arndt, in junior high, who encouraged me to enter science fair competitions. So, starting in winter 1960, work began on my second telescope. It would be a big one, and I planned to grind my own primary mirror.

It is useful here to repeat how supportive my parents were about education, particularly my mother. Once, while tracing maps on the dining room table, I pressed so hard that the image was permanently cut into the table surface. Other mothers would have been angry, but the marks were covered up and I was just told to be more careful.

The telescope construction involved a lot of such forgiveness. Dad helped with some of the woodwork for the tripod stand, and my grandfather Hans, a sheet-metal worker, soldered the metal tube. But I ground and polished the main mirror and built the optics.

My grinding stand was placed on the short connecting passage between the porch and our home's living room. Creating the right curve on the mirror involved a lot of grinding compounds and, in the final stages, a fine rouge powder for polishing. The grinding process stretched from March to August, with the bright red rouge powder slowly migrating from my stand to the adjacent living room furniture and carpet.

I melted tar/pitch on the kitchen stove to make the polishing surface and later melted lead there as well for the telescope counterweight. Dangerous vapors bubbled off the stove each time.

My parents may not have known how risky my actions were, as well as others I will describe later, but they trusted me to do the right thing. They let me make all kinds of things, and they let me make mistakes, big mistakes sometimes, and never came down on me. I was an experimenter. I am still an experimenter today because of their attitude.

The telescope was eventually finished and had excellent optics. The primary was aluminized in a small shop on Hennepin Avenue. I won a blue ribbon at the Twin Cities Regional Science Fair with a half-completed telescope and received a red (second place) ribbon at the State Science Fair in St. Cloud. The big blue telescope still works well and has been my companion ever since.

High School (Yahoo!)

I finally made it to tenth grade at John A. Johnson High School. Johnson was located only a few blocks from my junior high school on a bluff overlooking downtown St. Paul. It was also an old building and the atmosphere was a lot like that in the movie *Grease*. The students were into rock and roll, cars, dating, and having fun. I was pretty oblivious to all of that and settled in as mostly just an introverted student.

Early on we were tested and asked to think about our futures. I created notebooks comparing the careers of teachers, engineers, and scientists. I didn't really make any decisions then, but you see my drift.

We were also tested for athletic skills. On the shot put test, I had the weakest throw of the entire sophomore class, about 6 feet. By contrast, in the high jump, I cleared the highest level—4 feet, 10 inches—and decided to join the track team. Throughout high school I probably averaged about two hours of high jumping every school day. During the summer, I jumped at the Wilder Playground, dragging the standards and crossbar from the storage shed each time. Each day I had to sift through the sand and straw to pick out the broken glass and large rocks before starting. I wore a foam protector on my left elbow and knees but usually had allergy pimples and scratches there because that's where my weight would fall during the jump. My dreams and goals were updated in high school to wanting to win a gold medal at the Olympics, but it was not to be.

I also ran on the cross-country team and played on the basketball B-team, but there's not much of a story there. I should say, for the diligent runners in our family now, that I did cheat at cross-country running, often walking on the far side of the course, out of sight of the coaches. There were always a few kids from other teams willing to walk and chat with me. We would, of course, dash to the finish line with the coaches watching, but I think they knew what we were doing.

Chemistry

Mr. Westerlund's chemistry class was one of my favorites during my sophomore year. I didn't know much chemistry going in, so I learned a lot. Toward the end of the year, Mr. Westerlund did something that would probably get him arrested today. He knew I was collecting equipment to start a basement chemistry lab at home. I'd saved up money and purchased a slew of glassware and lab equipment. Dad had even tapped into the gas line and added a valve leading to my new Bunsen burner. He liked to solder pipes and do plumbing around the house. We found an old metal desk to do experiments on. The only things missing were actual chemicals. When the school year ended, Mr. Westerlund gave me 3-inch bottles of every chemical in the school lab, for four cents each.

Throughout the summer and the rest of my high school years, unsupervised chemistry experiments were taking place in my parents' basement. I did dozens of experiments, including one that filled the house with the smell of hydrogen sulfide. One time I etched my teeth by sucking too hard on a pipette filled with hydrochloric acid. I spit it out and rinsed my mouth quickly, but my teeth had a gritty surface for months.

The most memorable result from this time was my mistaking P (phosphorous) for K (potassium). The bottles were marked properly, but I had forgotten the symbols. My cousin Jimmy was with me at the time when I noticed that the bar marked "P" seemed to glow when I rubbed it—odd for potassium. There was too much light in the basement so we decided to go upstairs to my aunt and uncle's lower floor of the duplex and then into their hallway closet to see the glow better.

I held the bar with a burlap cloth and rubbed it more vigorously in the closet until it really started to glow. Then, unexpectedly, it ignited. Sparks were flying off in all directions so we ran out of the closet, threw the bar into the bathroom sink, and opened the faucet.

The hallway had filled with a thick white smoke, which I later learned was airborne phosphorus pentoxide, a poisonous powder, which with water can produce phosphoric acid. More ominously, my aunt's hallway carpet now had many 2-inch burn spots where the fiery sparks had landed. She put fans in several windows to blow out the "smoke."

Later, I heard rumors of negotiations between my father and uncle about this. My dad might even have purchased a new carpet for their hallway. But my parents never punished me. Again, I was simply told to be more careful.

Years later, with my Aunt Francis in a wheelchair shortly before her death, I asked her, "Do you remember the time I burned your carpet? Did I ever apologize to you for doing that?"

She looked me in the eyes and said, "Sure I remember. And no, you never did." I told her I was really sorry.

Mr. Westerlund did one more thing that was very important. He signed me up for summer Chemistry Camp, a two-month class at Augsburg College with some of the best students from around the Twin Cities. Professor Courtland Agre, whose son would later win a Nobel Prize in chemistry, taught the course. All I knew at the time was that he used a lot of words I had never heard before, like "cognizant." Also, the other students were studying and reading while eating lunch, something I had never considered before. I was suddenly competing against the children of lawyers, doctors, and truly "rich" people. And it wasn't going well. On the first test I scored 49th out of 50 students.

I compensated by studying harder than ever before. A small desk was set up for me in the dining room because I needed to work late and I was still sharing a bedroom with my sister. Mother often came out at 2 a.m. to suggest going to bed. By the end of the summer I was scoring first or second on the class tests, and I had learned what "college-level" meant.

Eddie and Janis

My father held jobs as a clerk or warehouse man through most of his life. During his work life at Conoco Oil's local office he had been paid consistently. We were lower-middle class, but secure. We had enough money to be the first to try out a few new products, like the tiny TV we had as kids in the '40s. I remember a half dozen little people huddled around the TV to watch *Kukla, Fran, and Ollie* or *Howdy Doody*. Occasionally he would bring a projector home from work. We'd set up all the house chairs like a little theater and pull down the shades. I charged a penny for neighborhood kids to come over and watch Dad's "movies," actually commercials from the Conoco Company.

Dad would sometimes repair TVs, checking and replacing the tubes in them to make a little extra money. He had worked two jobs once to make enough to buy me a great electric train set, which we lovingly set up on a 4-foot-by-8-foot plywood sheet in our attic.

Unfortunately, when he was about age 42, and I was 13, he lost his job. Our lives became much more difficult when he couldn't find a new one for a long time. Janis and I could soon see that it was eating away on him and the family in general. He and my mother argued more about the silliest things— how supper was cooked, whether it was done on time, and about small purchases. Sometimes they'd yell. After bedtime, I often heard them quarreling in the kitchen.

About this time, Janis got sick. She had been a vigorous, well-fed 10-year-old. I teased her, calling her "fat-seat." In fact I teased her a lot as we were growing up, especially when our parents weren't looking or nearby. That changed when she got sick. She started having stomach pains and losing weight. After a long time she was diagnosed as having Crohn's Disease. Throughout my high school and college years she underwent a series of operations to remove portions of her small intestine until there was little left. She became emaciated. When I graduated from high school, she weighed only 70 pounds.

So I can only imagine the pressure on my parents during those years. Dad eventually got a new job at Central Warehouse in the St. Paul Midway area and was later promoted to Router, the person who sent the trucks out and took the heat when someone was late. It was a tough, stressful job, but Dad always had a smile ready. The family balance returned to normal, but Janis remained ill for a long time, and the costs of two college educations loomed in the future.

Sports and Puberty

Each year I tried to participate in at least three sports—cross-country running, basketball, and track. The real athletes at Johnson played football, hockey, and baseball. I was clearly second tier.

Sports were fun, except for one thing—the showers. For some reason, perhaps my Scandinavian heritage or generally bad genes, I matured very slowly in high school. In the last two years I grew 12 inches in height, but for most of high school I had no hair in the "new places." In the showers, I was called baldy and worse. I thought something was wrong with me, that I was half-girl or something. Sleeping was difficult when thinking about the possibilities.

Around that time one of my best friends was Dan Peterson. I'd go to his house to play chess and sometimes we'd take the bus to downtown St. Paul to play pool at a large pool hall. We would talk politics—he was a big Goldwater fan. A couple decades later, I learned that Dan was homosexual. But in high school, for me, he was just a big, somewhat odd, interesting friend.

I still have a rather young-looking face and like to think the slow development in high school was simply a genetic quirk. I'm hoping that I'm on a slow-aging clock and will live longer than most others—at least to my deadline of 81.

Hepatitis and Mr. Abdella

In the autumn of my junior year, I had to drop out of cross-country due to fatigue, which was diagnosed later as hepatitis. My eyes had the characteristic yellow color and I became very sick, staying home for several months and missing out on basketball season. After awhile, I started getting visits from Mr. Abdella, my math teacher, who would drop off homework from all the teachers. No teacher in my life was as important as Mr. Abdella.

After the Russians launched Sputnik, there was a flurry of activity for the U.S. to catch up in space and the race to the moon, of course. But there were also important changes in U.S. research and schools. The National Defense Education Act affected education at all levels and in Minnesota led to the Developmental "D" class curriculums. High-potential kids with an interest in science and math had special classes: 9-D (biology), 10-D (chemistry), 11-D (physics), and 12-D (advanced studies in biology, chemistry, and physics), and similar enhanced math classes. Mr. Abdella taught the accelerated math programs for the 11th and 12th graders at Johnson.

Our classes were small. By senior year, only about 10 students were in the D math class. Abdella was a wonderful teacher—quiet, humble, and extremely effective. He always wore a dark suit and often would back up to the blackboard and get chalk all over his back. He was also funny. We loved him. Once the entire class hid in a side room with a small window to the math room and watched him look around totally befuddled when his class was missing. Then we burst out, yelling "Surprise!"

Mr. Abdella spoke with me at home when I was sick and told me to go as fast as possible on my own. Before returning to school, I had finished all the junior and senior year math material, including matrix algebra. So Mr. Abdella started bringing me his college math books, and I loved it. I also started doing practice tests for the National Mathematic Association's yearly competitions.

I developed a new theory and wrote a paper about how to work with the different infinities, using echoes of complex number theory. But one problem he asked especially motivated me: "What is i to the i power, where i is the imaginary constant?" The answer, by the way, is e to the power of $-\pi/2$, or about 0.20788.

Sometimes a teacher can change your life, and Mr. Abdella did that for me. I continued to see him occasionally for a couple years after high school, but then lost track. A few years ago I learned that he was in a hospital and possibly dying, so I went to visit. Upon entering his hospital room, I looked at him and simply said, "I'm Ron Peterson." His face lit up. His brother was there and we spent an hour talking. I told his brother what a great teacher Mr. Abdella was, and Mr. Abdella told his brother what a great student I was. He said there wouldn't be a miracle with his illness and that he would probably die soon. The next day he was gone.

I attended Mr. Abdella's funeral and thought of many things to comment about him while other people rose, one after another, to say kind things. I wanted to speak but couldn't somehow. But then his brother stood up, saying the day before Ed Abdella died a student had visited him to thank him for changing his life.

College Prep

Before my senior year our family moved to a new house at 1659 Chamber Street, only three blocks from Lake Phalen. I had to ride a city bus to school, but Janis and I finally had separate bedrooms. I got a nice small desk in my room so I could work late into the night.

I spent the summer before my senior year at Physics Camp at St. Cloud State University and completed a project to measure the air breakdown voltage in a high temperature furnace that I built. There was a lot of asbestos in the furnace, which I have apparently survived. Physics was really enjoyable, but I had no inkling that it would be my career. By that time I had decided to be a mathematician, preferably the best of the 20th century. I had started to think big. College was looming.

I had been selected to be President of the Johnson H.S. National Honor Society my senior year. But more oddly, by joining a weird group of friends who had created something called the Zarf party (sort of an anti-cheerleader and anti-good-looking-jock coalition), I had been elected to the Student Council as well. The Zarf leaders had organized dances, very frightening to me, and promotional materials that seemed to work.

I only applied to three colleges—the University of Minnesota, Hamline, and the California Institute of Technology or Caltech. The Massachusetts Institute of Technology, MIT, sent an application that seemed very long and difficult so I skipped it. My science and math teachers had encouraged me to apply to Caltech, which seemed like a long shot, but apparently a Johnson grad had been accepted there once before. Caltech really didn't seem to be financially possible, but I wanted to see if I could get accepted because it was the most selective university in the country.

I got a perfect score on the University of Minnesota math entrance test, which had never happened for a Johnson High School student. My SAT scores were good, about 760 in Math and 650 in English, but they were below average for Caltech. That winter, an interviewer from Caltech came to talk to my teachers and me. Mrs. Moosbrugger, my English teacher, gave the interviewer a copy of a paper I had written about how to tie your shoelaces, which included diagrams. She thought that would be best for a technical school. I have no idea what the other teachers said. The fact that Johnson was an undistinguished, inner-city school, with many kids hanging out on the street, I think helped influence the guy. Then began a long waiting period to find out where I'd go to college. But life didn't stop and my busy senior year of high school stretched ahead.

In November 1962, my grandparents, Ida and Hans, celebrated their 50th wedding anniversary with the whole family. As usual, the men, including me, played cards after eating, which was always interesting. My uncle Carl was a staunch Republican and my grandfather Hans was a union Democrat. Knowing I was going off to college soon, Carl pushed his views about competition and capitalism. He sometimes cheated at Canasta and said that was part of the game. I learned to hide my cards. My dad was neutral in the discussions about politics and Grandpa Hans wasn't as articulate as Carl, but I won't forget his final argument for me. Grandpa said, "Someday you will understand."

On a frigid January evening, with temperatures about 10 degrees below zero, a few students in advanced English and science classes were given a chance to demonstrate their work at a PTA meeting. The English students read poems and such. I was selected to do a science demonstration. I set up a chemistry experiment on top of an overhead projector, so everyone could see the change of color and bubbly reactions. Somehow the heat from the projector caused my experiment to go out of control and smoke began to rise. The smoke was magnified onto the large projector screen making it look terrible and dangerous. Some teachers looked horrified and moved forward to save me. I doused the experiment and made a quick exit from the stage. Years later, I learned that people who became very important in my life were there, and they told me they had never forgotten my dramatic demonstration.

I later overheard a few teachers talking about me, saying what a good science student I was, but then adding, "Too bad his English and grammar are so bad." That's the first time I understood that my dad's northern Minnesota-Norwegian syntax had apparently rubbed off on me. It was something I'd need to work on.

That spring, Hamline and the University of Minnesota accepted me, and surprisingly, Caltech did as well. The Caltech letter offered a $2,000 scholarship and an $800 loan. The total cost at Caltech at that time, including travel and room and board, was about $3,100—I was in. The beauty of the California Institute of Technology is that they are totally unbiased about applicants. If you get in, they don't care how much money your family has; they'll make it affordable. That is still true today, even though the actual real cost of one year is now about $180,000, and they ask wealthy people to pay around $60,000. If you are poor, you can get in, and the cost will be much less.

My entire senior year was consequential. Johnson won the Minnesota State Hockey Tournament in miraculous fashion (winning three games by only one goal). I was second in my class of 390. At the state track meet, I high-jumped about 6 feet, good for fourth place in the state. And on the National Math Competition I had tied for fifth place in Minnesota. In summary, I was a slightly better high jumper than mathematician.

Caltech or Bust

Entering Caltech was both the most joyous and terrifying experience of my young life. I had spent the summer getting ready, sort of. Many hot days were spent on a reclining lawn chair in my parents' backyard with one hand on a garden hose and the other on a book. I studied one book very carefully: *Calculus Made Simple*.

After two months of laziness, my father said, "You need to get a job." So for one month I unloaded turkeys from train boxcars (Butterball turkeys, three to a box) at Central Warehouse, where my dad worked. Seeing him at work helped me better understand his life, how he was the center of office actions, how he got calls regularly about problems with deliveries, and how he was always friendly under pressure. He made the coffee and brought donuts for the workers (I think people threw change in a basket for him).

My job paid well. I got stronger too—those turkeys roll around in their box and aren't light. The foreman there wouldn't let me calculate the number of boxes piled on each pallet. Perhaps he thought multiplying three numbers would be too difficult.

Before leaving for college, my dad said, "Maybe you should go to barber school some summer. I'm not so sure about this science thing." My mother's advice was simply, "Believe in God and come back to Minnesota when you are finished studying."

So off to college in California I went, for the first time on any airplane. This was back in an era where people smoked on the plane and the airlines handed out little cigarette packages as a gift to all the flyers. They also served real meals on china with metal forks! Ignoring all that, I stared out the window nearly the entire flight, fascinated by the towns, farms, and mountains passing under.

Landing at the L.A. airport frightened me. Everything was so new and huge. Eventually I found the baggage claim and picked up my small suitcase and my dad's Air Force duffel bag. Printed instructions from Caltech told me I next needed to find a shuttle to the Pasadena Huntington Hotel, which only ran once an hour. Riding along the Harbor and Pasadena freeways, I again stared out the window gaping. L.A. was very big, even bigger than Minneapolis and St. Paul combined.

The Huntington Hotel, one of the oldest (1907) and nicest hotels in the Los Angeles area, is named after the rail baron Henry Huntington. Many significant events in my life would take place in the Huntington, and I still return there often. It's located near the richest areas of Pasadena and San Marino, with rolling hills, beautiful gardens and mansions everywhere. To live there you still need to hire a gardener or complete special city gardening classes.

On my first day at Caltech, I sat in the imposing Huntington Hotel entrance portico waiting for my ride. Eventually a Caltech upperclassman arrived and threw my bags into the back of his open-top MG and we sped through the curvy streets, finally screeching into a Caltech parking lot. He showed me to my temporary room in Ricketts House, introduced me to Erik Lippa, and disappeared. Erik was also from Minnesota, so they must have figured we'd have things in common. A thin student from St. Louis Park, a Minneapolis suburb, Erik was urbane, occasionally smoked a pipe, and was friendly and self-assured. We settled in.

After recovering from the trip, I got my new books, walked around the town and campus, and started studying. I was afraid of flunking out. I was also having trouble breathing.

A couple days later, all the freshman were bused to orientation camp in the mountains, near Mount Baldy, which nearly killed me (I exaggerate only a little, as you'll see). We were each given bunks in unheated cabins and got to know one another. The new frosh played games, hiked in the woods, and were "oriented." The 160 frosh that year learned about the seven houses (not fraternities, not dorms, but something in between). We learned the school's history and school song. As an all-male school, we were told how to meet and properly treat girls.

At a final evening gathering, after a student talent show, we were told where we really stood. They said that we were similar to the top 160 people at MIT, except that each MIT class had 800 people who weren't as smart. They said one of every three frosh at the camp would flunk out eventually—so study hard. We were told that 99 of 100 engineers kept the world going, but the hundredth person, the Caltech grad, changed the world forever.

As all this went on, I got sicker and sicker. Before leaving Minnesota I'd had difficulty with asthma from ragweed and dust. The smoking on the airline, allergies from the new foliage in Pasadena, and finally cold nights in the mountains all conspired to give me pneumonia. After visiting the Caltech Student Health Center, I spent the next three days in bed. Writing to my parents, I said I was falling behind fast and was sure to flunk out, but I would try my best.

Not long after recovering from pneumonia it was Rotation Week. This was, and still is, a nerve-racking process, although it's set up so you can't be rejected by all the houses. It involved eating at a different student house each night and being interviewed. We tried to sell ourselves to each house. Each house claimed to be the best and to have a distinct personality. For example, Fleming House was clearly the jock house. Both Erik Lippa and I ended up in Lloyd House, one of the three newer buildings. I'm not sure what the Lloyd House personality was—I thought of it as the misfit house.

But Lloyd had pulled off the great Rose Bowl hoax of 1961, where the halftime flip-card show had been altered so that the word CALTECH appeared in bold, block letters for a 30-million-viewer TV audience. This bold prank involved intrigue, deception, and a sense of humor. It is still described on Wikipedia as the "greatest college prank of all time." So we did have that reputation to live up to. I think the Lloyd upperclassmen just liked the fact that I was a high jumper as much as anything. Erik and I continued as roommates.

Settling in at Caltech

When I mention that I went to Caltech, everyone first wants to know about Richard Feynman. Physics was the queen of sciences in the 1960s, and Feynman was the king. He'd worked on the Manhattan Project, had revolutionized quantum electrodynamics, and he had taught and published a new treatment for first and second year college physics—the famous "Feynman Lectures." Unfortunately, he had finished teaching the course the year before I arrived. His colleagues continued to teach the course and Feynman did come in to give some special lectures on how the eye worked, about lightning bolts, and how he had derived Newton's Laws of Planetary Motion from first principles without calculus. In his books and lectures he mixed science with his philosophies of life. As an introvert, two tenets I absorbed, which probably changed my life, were these:

- Don't care so much about what others think of you.
- Generally, don't trust authority figures or institutions.

He was an amazing character. He never just entered a room; he waited until everyone had settled down and then blasted into the lecture hall, door exploding, talking to someone or himself very loudly, marching down the stairs like he was defying the gods, long hair flying in all directions. Once in front, he would often start with an outrageous statement that gripped every person in the room. He defined the word "charismatic."

Later in his career, he would introduce the concepts of nanotechnology and quantum computing. He would work on the Roger's Commission investigating the shuttle explosion, and he would win a Nobel Prize in 1965. In 1999, he was ranked as one of the ten greatest physicists of all time.

There were other Nobel Prize winners walking around Caltech. Linus Pauling (two-time winner for his work on the nature of the chemical bond and his successful efforts to end atmospheric nuclear bomb testing) also lectured to the freshman class occasionally. To date, Caltech faculty or grads have won 38 Nobel Prizes, not bad for a small school with only 200 students in each class per year.

I did well my first year in physics, chemistry, and math, averaging As in all those classes. Things were not as first-rate in English or history. While Caltech emphasizes the sciences, it is also quite strong in the humanities, and student scores then and now are almost as high in that area. The competition is tough.

Students called my first year English professor, Harvey Eagleson, "Easy-A Eagleson." He attempted to turn his Caltech students into culturally oriented, verbally adroit men of the world. Our assignment each week, in addition to reading one book, was to write 500 words on a subject of our choosing. He loved one of my early stories, which I wrote in a style like the confessions of a drug addict. He read it aloud to the small class without comment. The story was about becoming hooked and slowly being unable to resist cravings and obsessions. At the end, the story revealed that the addiction was to Mounds candy bars. The entire class groaned and then sneered at me as he handed the paper back.

Soon after, I started putting off writing my little essay each week and eventually got about eight behind. Perhaps I didn't know how to keep up the quality. More likely, I was lazy or too interested in the science classes. Once behind, it seemed hopeless to catch up. In the end, Eagleson gave me an E for incomplete and forced me to write a 3,000-word essay to get the grade moved up to D-. In his class a C was considered a reprimand. My work in history class was almost as abysmal. I tried a different, more focused approach toward the humanities my sophomore year, which helped a little.

Life in the Student Houses

Lloyd House life was a revelation and quite unsettling. Bombarded with new ideas and ways of living, I probably learned more the first year at Caltech than at any time after the age of three. In the short, second floor hallway where Erik and I lived, known as the Virgin Islands, our room became a hangout and a new world. Steve Abramson, Gary Berman, and many others came to chat in our room, some assertive and worldly East and West Coast types. Future good friends, Ernest S. K. Ma and quiet Tom Buckholtz, lived in nearby alleys. They were brilliant and later would lead interesting lives. As just a typical example, Erik Lippa graduated with top grades at Caltech, got his doctorate in mathematics, then decided to be a medical doctor, and later worked as VP of Global Medical and Statistical Services at Allergan.

We freshmen talked about everything, including religion. I told them I was Lutheran and they left me alone, saying that it sounded like a wishy-washy religion. I admit that I was a coward about religion. Church services and our pastor in Minnesota had seemed rather boring. So I essentially went underground about God and didn't talk about religion or see the inside of a church again for nine or ten years.

I also learned about debate, culture, and music. My family had never really talked about politics or history at the dinner table. We worried more about daily life. Similarly, in high school, male students were into sports, cars, jobs, classwork, and what other kids were doing, not philosophy.

At Caltech, I spent a lot of time listening. My fellow students were sophisticated. They played every sort of music. Many large speakers were pointed out windows playing classical music, which I had barely heard before. On a small tape deck, I began to record everything I liked by borrowing music from others. I remember the exact day I first heard infectious Beatles music. "I Want to Hold Your Hand" was playing loudly as I ambled down the shady Olive Walk, the main path to all the labs and classrooms. And every "techer" through the years remembers Wagner's "The Ride of the Valkyries" playing out the house windows at 7 a.m. during midterm and finals weeks, coming in wave after haunting wave of music, to be sure everyone woke up for the exams. By the way, many exams at Caltech were take home, some even open book. The Honor Code was an important part of Caltech life. It was assumed that no one would cheat on exams, ever.

I also learned about alcohol. Although officially prohibited, drinking did occur on campus, especially on weekends. I was introduced to beer for the first time at a keg party. The keg was set up in the shower and students repeatedly went in to fill up. Another memorable occasion for me was an evening when many were drinking in a circle in someone's room. They gave me a large glass with a clear liquid that had very little taste. Later, after finishing off that tumbler of gin, I was photographed, prone on the floor, "swimming" back to my room, unable to walk. I never drank consistently, really hating the hangover part.

Early on in Lloyd House, the freshmen were given a "Purity Test," 50 questions about sex of increasingly bizarre forms. Since I had yet to kiss a girl and my father had never told me about the birds and bees, I scored a 3, which might have been a record low. The terms used in the test were mostly a mystery to me, although some were explained later when I asked.

I almost asked a girl on a date in high school. I heard Nancy Frank saying she had no one for prom night at Johnson. Blurting out, "I'll do it," turned out to be a mistake. My mother forced me to call Nancy to apologize, and not just fail to show up. Kids in Minnesota in 1963 could be rather innocent, and I was about two standard deviations below the Minnesota norm.

California girls were on another plane entirely. Since Caltech was an all-male school, women were rarely on campus. But on some weekends, girls came from nearby colleges, like Scripps, Pomona, Occidental, or even high schools, for dances. Groups of Caltech guys would stand around and rate the beauty or sexiness of each as they walked in, not unlike an auction.

Once, when a group from Scripps Women's College in Claremont arrived, I actually dared to dance. Later, sitting on one of the couches in the lounge, the girl I'd danced with put her leg over my thigh unexpectedly as we talked. I'll never forget the effect it had on me.

After Christmas/Winter Break I flew back to California for $90 round trip (student stand-by, red eye). While deep snow still covered Minnesota, it was spring at Caltech, puddles on the ground and time for track to begin. Still weighing only about 150 pounds and 6 feet tall, I was very bouncy. At the end of track season, I jumped 6 feet, 4 inches and tied for first among freshmen in the Southern California Small College finals. I got the medal after a flip of a coin.

Sometime that winter trimester I decided to switch my major from mathematics to physics. One Lloyd friend kept happily talking about how being a mathematician meant working on very abstract ideas and never having to do anything very useful in your life. Thinking about my father's words, how he wasn't so sure about earning a living with the "science thing," made me cautious. I did want to do "useful" work.

Erik Lippa and I made a big mistake that spring by storing used soft drink bottles on our window ledge, stacking them up in a giant pyramid. Each room had a large picture window and in our case it faced the Olive Walk that everyone used. Someone bounced a tennis ball off the window from outside and dozens of bottles crashed to the concrete floor. We were finding glass fragments for weeks.

For spring break, Tom Buckholtz asked if I wanted to stay at his parents' house in Palos Verdes Estates for the week. I said yes and also agreed to be his roommate the following year. Tom, the studious guy down the hall in Lloyd House, is probably the most competitive person I have ever known. He was eventually first singles on the Caltech tennis team. Each morning during the visit, Tom, his father, and I would go out at about 5:30 a.m. to play tennis after running two laps on a local high school track. Tom's father, Joe, an aerospace manufacturing executive and community leader, was equally competitive, blasting the tennis balls back with all his strength. At every meal, he and Tom competed at speed eating. Tom's mother, Sylvia, ate slowly, explaining that it would be impolite for her guest to be eating alone. Tom ultimately got his doctorate in physics from Berkeley, led a company-wide innovation program at Pacific Gas and Electric, and was Co-Chief Information Officer for the Executive Branch of the U.S. federal government. He is now trying to solve some of physics' greatest problems using a new quantum math he has developed.

As freshman year came to a close, my experience with women grew. Despite my resolution not to get distracted from my challenging studies, one girl I'd met started leaving cookies in my dorm mailbox each week. She was a high school girl from Altadena who I called the "Cookie Girl." I had to leave a note for her in my mailbox so she would stop. Another young woman from St. Mary's College in Hollywood wasn't very good looking, but I danced most of one evening with her. On a car ride back to her dorm, she said, "You probably won't call me, will you?" I lied so she wouldn't feel bad.

But my most serious "girlfriend" was a junior at Arcadia High School named Laura. At one of the exchanges I danced very close to her, felt her breasts on my chest, and generally rubbed bodies. While on Caltech's Olive Walk later that night, she became the first girl I ever kissed. The Olive Walk is a picturesque paved path leading through the student houses toward the classrooms. At night, it's a romantic dreamy stroll, although you do need to avoid squashing fallen olives.

We had an amazing date a couple weeks later. With about six people packed in a car, we rode to Hollywood to see a first-run movie. Afterward, we ate at a restaurant that played old-time songs (like "Camptown Races," "I've Been Working on a Railroad," etc.) and I sang along. Laura was surprised that I knew all the songs and I was surprised that she didn't. When we all got back to the car for the 40-minute ride home, we found it wouldn't start. So we all pushed the car about 10 blocks down Hollywood Boulevard, past Grauman's Chinese Theater to a gas station that was closed. We called our dates' parents to say we'd be a little late. By the time someone picked us up and then took everyone home, it was about 3 a.m. This probably looked bad: a college guy out with a high school junior until the middle of the night was definitely bad form in 1964. I thought I'd seen the last of Laura.

But Laura asked a friend to pass the word to me that she would very much like to go to the Caltech "Lost Weekend." Lost Weekend started with a formal (prom-like) dance at the Huntington Hotel ballroom. Laura's hair was bundled on top of her head in a very fancy way and it smelled great. I remember kissing her the entire last dance. We strolled in the gardens behind the hotel on a beautiful evening under the trellises and moonlight. The night was magical and very romantic.

On Saturday, everyone took a bus to Long Beach harbor and rode a boat to Catalina Island. I got seasick on the 26-mile ride, but hiking to the top of the island hills and later lounging in swimsuits on the beach were both wonderful. The final Lost Weekend event was breakfast together on Sunday morning at a nice restaurant and then visiting the Arcadia Arboretum. On the drive back to her parents' house I accidentally brushed Laura's breast and she quickly pushed my hand away. I wasn't going to get another point on the Purity Test easily. It was almost time for summer break, and we both promised to write.

Other than finals, only one important event remained during my freshman year—Ditch Day, a secret date when all the seniors disappear for a day of fun, often on a beach. They set various puzzles, blocks, or locks on their rooms and the freshman are challenged to somehow crack the puzzles and let the seniors know they failed by changing their rooms. Over the years, many amazing things have happened. Rooms have been filled with water with goldfish swimming around. A Model-T car was completely disassembled and reassembled and left running in one room. Student rooms have been coated with asphalt and made to look like a highway.

The puzzles have been equally creative, requiring voltages, heat, cold, and intricate combinations to pop the lock, often with tricky instructions attached to the door. In 1964, all the doors had regular locks that every freshman learned to pick in the first weeks on campus. Those locks have since been replaced with electronic combination locks, but during my freshman year most seniors had clever mechanical devices on their doors to keep people out. They also had mechanisms, like pendulums, to break the large picture window if anyone tried to get in through the crank-out side windows. My Lloyd House freshman class had a big problem; 15 seniors still lived in our building. That's a lot of clever puzzles to solve in one day.

We got right to work as soon as someone noticed the seniors were gone. Dangling a small mirror in through the ceiling light fixtures from the attic enabled us to crack into three rooms after seeing the mechanism they had used. But 12 rooms still remained—it was time for brute force. The large bottom drawer in some built-in room cabinets were pulled out and the wall structure behind (plaster and wire mesh) were cut out and bent back. This enabled us to push out the similar drawer in the senior's room. Being the skinniest freshman, my job was to squeeze through the resulting hole to enter the senior's room. We got the job done and altered the lock mechanisms so the seniors would have an almost impossible job getting back in. Over the summer, the Buildings and Grounds people left a note asking that we avoid permanent building damage on the next Ditch Day.

Freshman year was over. I rode home with two other Minnesota guys in just 40 hours of continuous driving. I had the 3 a.m. to mid-morning shift and got to see the sunrise in New Mexico before falling asleep in the backseat. It was a new kind of sunrise for me, with no trees and a beautiful bright salmon color glowing on the mountains. Freshman year at Caltech had changed me—at least part way—to becoming an adult.

Caltech Year Two (1964–65)

The summer of 1964 I got a job working for 3M Company. During the first two weeks they didn't know what to do with me, so I was given huge buckets of screws and told to sort them. Eight hours staring at screws. Each night I dreamt of screws. Just shutting my eyes, I saw screws.

Eventually they started having me work with people in the inspection department. Precision instruments were used to measure incoming parts for 3M products. Only a specific fraction of the hundreds of parts were checked to approve their vendor lots. I learned about statistical analysis. The job became mildly interesting, much better than sorting screws, and it paid well.

Getting a summer job was important to reduce the loan amount needed for Caltech. All the loans at that time were "National Defense Loans," which meant they would need to be repaid after my last year of education at four percent interest. Student loans today are much more punitive.

Of course, I didn't actually write to Laura despite our promises and she never wrote to me. Back in Minnesota, the memory of dating her seemed to fade. I was much more passionate about science and learning. And Laura, in retrospect, seemed pretty aggressive. My Aunt Ruby said there was plenty of time for girls after I graduated. And perhaps I was a little scared of Laura, by sex, and by what dating could lead to.

Caltech changed over the summer. The experience of our freshman class worried the school administration. The freshman work and competition had been intimidating. Out of the 160 freshmen who entered, only 90 returned for year two. That was not an acceptable dropout rate. So the freshman in 1964 and all subsequent classes had pass-fail courses their first year. I think the idea to open enrollment to women also began about that time, although it wasn't implemented until 1970.

I called Laura once when back on campus, apologized for not writing and said I wanted to see her, but I never contacted her again. A year later I learned that barely out of high school, she was married and pregnant. Perhaps I had avoided a big change in my life.

To get better grades, I vowed to work twice as hard in my humanity classes, which were still required in the sophomore year. The strategy worked, and my overall grade average crept up to B+ and then A-. But then the science and math started to slip, and once you get behind there, it's very hard to catch up.

Spending too much time in the Lloyd House lounge didn't help. I learned to play bridge and became fairly expert at poker. My poker strategy usually involved being very tight at betting most of the night and then bluffing big time when everyone thought I was cheap. That seemed to work very well with new people. A lot of extra cash was available to me, which I used to buy 24-bar boxes of Mounds candy bars from the vending machine man. I really was addicted to Mounds bars and wrote a letter to my mother promising to cut down to only two per day.

New friends appeared. One was a freshman named Steve Landy. He invited me to his parents' house in Santa Monica for Thanksgiving, which turned out to be quite interesting. With his entire family seated around their dinner table, the glassware in cabinets started to rattle and then a chandelier hanging over the table began to swing back and forth. The whole experience was alarming to me, but none of the others at the table seemed overly concerned. Steve's father said, "I'll give Mom 30 seconds to call." Soon their phone rang and Steve's father picked up and said, "Yeah, Ma. That was an earthquake." Steve Landy would become, perhaps, the closest friend I would ever have, before I met my wife.

Throughout my sophomore year I waited tables in the Lloyd House dining room, again to make money and reduce my loans. I learned how to balance five plates of food on one arm and serve food properly. The meals were very organized affairs, often requiring everyone to wear their Lloyd House sport coats, especially if guests were invited. In our case, those coats were puke yellow with a giant script L on the front. Lloyd House's officers presided over the dinners, made announcements, organized dances, and athletics teams, etc.

Respect between the Caltech students was a curious concept. Of course grades were relevant, but I think I received more respect for my poker skills, which my father had taught me.

Everyone was expected to have mechanical and electronic expertise; projects were constantly being built and experimented on around the dorms. Each house had access to machine shops and tools and surprising things happened all the time, like finding a floor-to-ceiling concrete wall appear in a hallway where none previously existed. Sometimes doors to student rooms suddenly disappeared, having been sheet-rocked and painted over to match the other walls. Chemistry and physics experiments occurred regularly in the open courtyards and people lived in the steam tunnels. (See the movie *Real Genius* for a detailed look.)

I think the peak of the respect heap was reserved not for those with the best grades, but for those with top grades who never went to their classes. I knew one such person, a year older, who lived down the hall from my room. Ernest S. K. Ma would average an A+ barely cracking a book. The books he did read weren't part of any class. He must have known most of the Caltech curriculum before he arrived. In his room, we would discuss primitive societies as he read Sir James Frazer's *The Golden Bough*, a comparative study of magic and religion. I heard the cantata *Carmina Burana* by Carl Orff for the first time in Ernest's room and grew to love it. He ultimately became a professor at the University of Hawaii–Manoa working in theoretical particle physics, and is now Emeritus Professor at the University of California, Riverside.

I went home for Christmas/Winter Break, as I did every year. It was great to see snow and family and exchange gifts. Starting about that time I began giving my Aunt Ruby a strange candle each year, and they became increasingly bizarre as the years passed. There were dragon candles, chickens, insects, turtles, eggs, and hippos. She soon started returning the favor, and I have dozens still displayed in our basement.

My flight back to Pasadena was always too late to see the Rose Parade, but our football team did play its home games in the Rose Bowl. The Caltech football team story needs to be told. At that time, the Caltech student section would fill one row of one section of the Rose Bowl, which can house 100,000 people. We had male cheerleaders, of course, on the field and the team was made up of great students but very few real football athletes. Many

on the Caltech team had never played in high school and the coaches tried
to make up the difference with intricate trick plays. That didn't work very
well. Caltech played pretty strong teams at the time, like Cal Lutheran,
Riverside, Occidental, Pomona, etc., and winning one game was considered
a good season. A giant bonfire on a campus street would follow any win. In
1963–4, we won two games and it snowed in Pasadena, both very rare.

One game was especially memorable. We played Cal Lutheran, which
was ranked among the top ten small college teams in the nation. Those of us
at the Rose Bowl cheered for Caltech for a little while and then went over
to the large Cal Lutheran side to sit with them. Cal Lutheran had girls in
the stands. But it was an unusual game because Caltech scored twice. The
final score was Cal Lutheran–33 and Caltech–9. Because of the relatively
close score, Cal Lutheran dropped out of the top ten rankings and never got
back in.

The houses played intramural sports as well, and I was a defensive line-
man on Lloyd's flag football team and actually fairly good. Due to my high
jumping and strong legs, I was able to abruptly cut and leave big defenders
looking, thereby getting to the quarterback. I did break my glasses when a
football hit me between the eyes, and I had to tape the parts together for a
trimester. This gave me a very nerdy look. I also twisted an ankle for the first
time while playing flag football.

I gained weight sophomore year and ballooned up to 180 pounds, which
ruined my high jumping for two years. As a senior my muscles had finally
adjusted to the weight and I was jumping 6 feet, 4 inches again, my all-time
record. The Fosbury Flop would not be invented until 1968. It could have
given me another four inches or so. Too bad.

For spring break, Steve Landy, a couple of freshmen, and I decided to visit
Death Valley. One freshman's father was a U.S. congressman and, more importantly, had bought a car for his son. So off we went, camping out at state
parks, enjoying the bleak scenery. For the ride back to L.A. we decided to
take a small road, a shortcut barely visible on our map, across the mountains.
There was a lovely view of the valley but forecasts of bad weather ahead.
While driving along the mountain road, it began to snow and our car slid
off an edge into a ditch. Somehow we got people to pull us back onto the
road with a chain hitch. We had passed the last village on the way up, so we

were lucky anyone saw us. We all voted to ride back down into the valley and home on a freeway.

As the second year at Caltech came to a close, I struggled in Feynman's quantum mechanics and had trouble with advanced calculus. Humanities were fine, but the slip in my favorite courses frightened me. Poker and goofing off would have to cease.

Caltech Year Three (1965–66)

Before going back to Pasadena for my junior year, I needed another summer job, and because Caltech's year ran into June, the pickings were thin. I used the Civil Service System to get a job at the Air Force Reserve Base at the Twin Cities Airport. Each day I rode the bus to downtown St. Paul and transferred to the airport. The work was in a shed attached to a hangar where C47s were brought for repairs. The job was to repaint the big carriers and sometimes, on smaller planes, to apply aircraft dope to cloth stretched over wings. Aircraft dope is a lacquer that tightens and stiffens fabric stretched over airframes, rendering them airtight and weatherproof. In high concentrations, addicts looking for a buzz sometimes sniff it. Since the workers in the shed spent long days in these vapors, they were all a little strange. It wasn't until my last weeks on the job that they let me use the paint-can shaker, because it was "dangerous" they said. On my final day, I was taught how to use the high-pressure paint sprayer on the C47s, also a high-tech job by their standards.

Many days I brought my father's chemical warfare mask, left over from WWII, to wear in the shed when the fumes were really thick. My fellow workers thought I was odd. Occasionally, I left the gas mask on while waiting outside on the curb to catch the bus home. I still had asthma, and the bus stop was next to a large field of ragweed. Cars stopped to see if I was all right, the drivers yelling out their windows. I guess a guy sitting on the curb wearing a gas mask may have looked a little weird. "No, go on," I told them, "I'm just resting." Today they would probably think I was a terrorist.

Back at Caltech for my junior year, my reliable participation on Lloyd House intramural teams the previous spring paid off—I was elected Athletics Manager, which became an important learning experience for me, sharpening my organizational skills. Early on, I interviewed all the new freshman and upperclassmen to categorize their athletic abilities. This went on a giant chart (about 4 by 5 feet) with dozens of sports on the columns and about 80 Lloyd House member names on the rows. For example, I knew which people were great at ping-pong, skeet shooting, and that one senior was a top-three high school wrestler in the state of Oregon. This was well before Excel spreadsheets had been invented.

The information was very important for the Caltech Discobolus athletic competitions. A Discobolus trophy changes hands several times a year when a challenging House successfully defeats the House holding the trophy. The challenger names three sports, and the House holding the trophy must accept the challenge in one. With my giant chart, we were masters. After successfully challenging once, Lloyd House held the trophy the rest of the year. We also did very well in regular intramural sports, winning the Interhouse Trophy at the end of the year.

Perhaps because of our sports success, the Lloyd House president made me co-chair of the Interhouse Dance event and promised to get me a spectacular hot date. Soon we were building "Paris" in our courtyard. Interhouse constructions are not a small thing; they often take two months of planning and two weeks for construction. They can involve flooding an entire courtyard and building boats, erecting caves throughout building hallways, fireworks, or a million other surprising things. They always have a theme, and in 1965 people came from many colleges to participate. This all ended years later when someone was stabbed by gate crashers. But in 1965, Lloyd House built Paris!

We scanned many books to design typical French houses and apartments and constructed half-size replicas in the courtyard, with lights in the windows to show they were occupied. There was an outdoor café and French waiters serving drinks. The courtyard had an Arc de Triomphe and a 20-foot Eiffel Tower was perched on the house roof.

Before we could pull this off, a near disaster occurred. Rain was forecast for Pasadena, which would have destroyed many of our outdoor constructions. So I went into action, rallying all freshmen one night to string sheet after plastic sheet, completely covering the large Lloyd courtyard. All the frosh wastebaskets were commandeered and filled with water to weigh down the plastic edges. It did rain hard that night before the dance, but our structures and decorations were saved.

A couple weeks before the Interhouse Dance, I was shoved into a phone booth by four or five Lloydies and forced to call a girl—the hot date I'd been promised. They wouldn't let me out, and I took a lot of time to work up my courage. The girl was Karen McKinney, music major at Occidental College, the kind of person skilled enough to play with orchestras. Karen was beautiful, and I was quite happy when we picked her up for the dance. She was very intelligent and thoughtful; I soon fell in love with her. We had a great time at the dance, and I actually called her again later without being forced. She was marriage material, but made it very clear on a subsequent date that she had no such intentions, for a long time at least. Karen would become a bone of contention with a future roommate of mine, Douglas Osheroff.

There really wasn't much time for females, however. The math and science courses were becoming very serious: Advanced Quantum Mechanics, Differential Equations, Spectroscopy, Topics in Advanced Algebra (much worse than it sounds), and elective courses like Semiconductor Electronics, Thermodynamics, and Analysis of Numerical Methods (computer science before there were computers). Humanities electives were still strongly encouraged. I especially liked a Psychology of Child Development course, that I applied often years later with my two daughters.

During autumn of year three at Caltech I would often study at Dabney Hall to get away from the noise around the student houses. Dabney was a very old but elegant humanities building and I could spread out my books and papers on a table while alone in a classroom, leaving the large swinging windows open to let the warm evening Pasadena breeze into the room. But then my Dabney Hall horror began.

Remember the scenes from the book or movie *1984*, where the hero, Winston Smith, faces his greatest fear, voracious rats in a cage inches from his face? Winston is broken; the defiant hero is brainwashed and lost forever. In my case, my greatest fear was, and still is, moths. One night, the warm lights and open windows of my Dabney study room attracted dozens of moths. With a folded paper I chased them and swatted, trying to kill them or force them out the windows. The moths won. After one swing, a large moth flew into my gaping mouth to the back of my throat. I tried to cough it out, but its horrible powdered wings were stuck. Finally, I just swallowed it. I kept spitting all the way back to Lloyd House and never opened the windows at Dabney Hall again.

My grades did perk up junior year, but only because of extraordinarily serious and diligent study. No more poker or bridge. There were only two diversions, *Star Trek* and *Batman*. Both TV shows started about that time and became the only shows where large crowds filled the Lloyd House lounge once a week. This ultimately led to the great Batmobile fiasco.

A local radio station ran a contest in which the prizes were a new color TV and, more importantly, a ride in the famous Batmobile. Since it was a numbers quiz, a bunch of us thought we could win. It was like guessing a lottery number, with clues every day about the various numbers. You had to send in a postcard with your guess. Well, we applied all our algebraic and superior logic to narrow the possible answers to only around 1,300. So with adjustable rubber stamps, a small group stayed up all night filling out blank postcards with all possible answers. We finished before dawn, drove the cards to the radio station, and stuffed them into the station's mailbox to avoid having to buy postage stamps.

Later that day, the station broadcasted a "clarification." They had said, "One number is repeated twice," which we, of course, interpreted to mean that a specific number appeared three times. But that wasn't what they meant. Their bad grammar had done us in. There were still thousands of possible answers, and later that week someone else won the ride in the Batmobile.

Steve Landy and I played a lot of tennis that spring and became good friends. Our manners of speech and modest personalities were similar. Steve and I helped Lloyd win the interhouse track competition, with me running high hurdles, pole vaulting, and high jumping. I taught Steve to high jump and with his very strong legs, we did well. Actually, I was a bit jealous of how well he did without years of practice. Much later, even now, when we write letters to each other, we sign them with the letters TOMOTLHIHHJT, "The other member of the Lloyd House interhouse high-jumping team." Don't tell anyone about this signature code; it's our secret.

Before leaving Caltech my junior year, I had a weekend of true decadence. Andy McKay's father invited him to enjoy his empty house, which was located only a few blocks from the beach in San Diego. Andy, a fellow junior, invited Steve Abramson, Joe Manke, and me to take a train south to stay in the house. I didn't really know what I had signed up for.

We took a bus to downtown Los Angeles and had a long wait for our train, so I was introduced to a strip joint porn show near the train station. My morals were falling fast. The train ride was fun, but as soon as we got to San Diego they wanted to stop at a liquor store. Andy had turned 21 so we were able to load up with many bottles. I learned about screwdrivers that weekend. All I remember about our first night is that I became very woozy, and as we prepared for bed, exceedingly late, I asked if I could sleep between two of them on a giant bed, reasoning that it might stop the world from spinning around. They laughed and teased me as I fell asleep.

The next morning was awful. We walked down to the beach and I sweated off the hangover all morning. Then, Saturday night, the drinking started all over again. Somehow we made it back to Caltech. There may be a moral here, but I'm not sure what it is. A few weeks later I was again safe in Minnesota with my parents.

Caltech Year Four (1966–67)

That summer I got a decent job at a place called the U.S. Bureau of Mines. It was one of about seven national labs dedicated to support the mining industry. The Minnesota facility was located on the banks of the Mississippi River under a flight path near the Minneapolis-St. Paul International Airport. At last I would be able to perform real science during a summer job. One of the directors, Dr. Kalafalla, was very excited to get a young Caltech student for the summer.

My grandfather had heard about my asthma and problems taking a bus to the airport area the previous summer. He was about 80 years old and not driving much, so he let me use his old car for the summer, a '57 Chevy. My commute became a daily adventure because the car used at least one quart of oil on every trip to work. I had to bring a second quart to add before driving home from the airport area. Every time I accelerated, a great cloud of black smoke filled the air behind the car. Being arrested seemed a certainty. But somehow I made it through the summer without being pulled over or the car breaking down.

At the Bureau of Mines, they did everything imaginable to rocks. One room had a powerful laser (a relatively new invention) to bore holes in rocks. In another room, which had three wire screen layers on all sides, a strong radio frequency (RF) field was placed on paddles on either side of oil shale chunks. I think they were trying to get oil to ooze out of the rocks, but mostly the shale just heated up and then cracked or exploded. Apparently they had originally done it without the screen room and it put out so much radiation that the airplanes flying overhead lost their communications. By the time I got there, they checked with the airport before turning on the RF generator to be sure no planes were overhead, even with the screen room trapping the radiation.

I was assigned to work with a long-time researcher who measured the heat capacity of sample rocks in a big machine. The rocks were heated up and then dropped into a chamber to see how much energy they had stored. Many temperature measurements were made at precise times to complete each experiment. My job was to make these measurements, plot the changing temperatures, and calculate the heat capacity of the sample. It was better than counting screws, but after a few weeks I got bored.

So during lunch, and at every opportunity, I learned how to program what passed as a "computer" in 1966, trying to automate my job. Soon I was shirking my assignment and working on the computer problem most of the time. My boss complained, telling Dr. Kalafalla that I wasn't doing my job. Amazingly, Dr. Kalafalla told him to leave me alone, to let me figure out how to best use my time. Soon another student was measuring rock temperatures and I was feeding his data into my new program. In those days, computer instructions were cut as holes in the narrow paper tape of a teletype machine. As the tape scrolled through the machine, numbers were added or multiplied as you instructed. So the measured temperatures and times were typed in and the machine popped out the heat capacity without drawings. My program was a success, simplifying the job, but I had been sort of fired by my boss. Soon I was assigned to the oil shale cracking team, which was more fun anyway.

Back at Caltech for my senior year, my new roommate, Doug Osheroff, and I decided to live off campus. We found a place about two miles from campus at the corner of El Molino and East Villa. An elderly couple lived on the first floor and we had the entire second floor, which even had a nice porch. Every day we could ride bikes down the steep hills of Pasadena to our classes.

For only $200 we completely furnished our house. We bought two beds, a sofa, chairs, lamps, cabinets, and a nice dining room table from "Sy's Unclaimed Freight." Sy guaranteed that the furniture was unused, if slightly damaged. Later we found Tinker Toys under the cushions of the sofa, so Sy might not have been entirely truthful.

Doug and I quickly settled in, studying diligently on the dining room table or at desks in our rooms. Doug studied very hard. Although at the time we seemed about equally talented, his hard work paid off big time later. I still claim that I was one of the people who convinced him to major in physics, rather than electrical engineering as he had originally planned, although Doug now denies my version. We were both taking a lot of physics courses by that time, including many grad-level courses.

One particular class involved a direct confrontation between Richard Feynman and me. A graduate course, Quantum Electrodynamics, became particularly challenging after a few weeks, and I worried about my likely grade. So, on the final day for dropping courses, many students who had signed up for the class just to say that they'd had Feynman as a teacher, were all lining up to get him to sign the drop form. The queue was very long, and I was at the end. As student after student stood in front of him, he would shout out in a loud voice, "No. I won't sign. Be bold, young man. Learning matters. Grades don't matter. Come back next Monday and try harder." Many of these students were chemists or biology majors, and he refused to let them bail out. Finally I stood in front of him, expecting to be blasted. I told him I was taking three other physics courses and had started to do research work with another professor, and I couldn't do his class justice. Apparently that was a good excuse. He said, "Okay, then," and signed my card. Too bad. The next Monday he announced that he was making the class pass-fail and that everyone would be passing. I ended up going to most of the classes anyway—for no credit.

A professor had hired me to do lab research that year, mostly constructing electronic circuits and boxes. His research involved magnetic resonance and I built something called a marginal oscillator. These were the key elements in what would later become Magnetic Resonance Imaging, MRI. Of course, today MRI is an important diagnostic tool in hospitals, using very expensive large machines and fast computers. I don't know if our research contributed at all to practical MRI technology, but it was interesting and I got paid.

Roommate Doug had an even better job, working up on Mount Wilson on an infrared sky survey being done by Professor Gerry Neugebauer. Neugebauer had built a large infrared telescope in a most creative way. A large cylindrical vat of epoxy was rotated about its axis until the epoxy hardened. Liquid in a spinning container takes on a parabolic shape, perfect for telescopes. So this was an inexpensive way to make a very large telescope mirror. Neugebauer is widely recognized as the father of infrared astronomy.

I performed one experiment at our house on East Villa. Every morning I had begun leaving food on our porch railing for the blue jays that lived in the nearby trees. Eventually I started putting bread pieces on a large oven mitt outside, and soon I could hold the mitt as they ate. The blue jays seemed to love the extra food. Leaving a little path of bread on the porch, they followed the path in whatever shape I made. One day I left the front door open. I wanted to see if they would go into our living room and then back out. It worked. After several days, they would peck their way in, 20 or 25 feet, past the living room and to the hall bathroom, always flying back straight through the front door. Then, one day, a bird got lost in our living room and started flying around in a panic. It took a long time, with all the windows open, to get him to fly out. The experiment ended that day.

Caltech was always able to attract interesting famous people to campus. My senior year, Robert Kennedy came to give a speech on the mall in front of Beckman Auditorium. Also that year, a concert was planned with Igor Stravinsky, the famous composer. By this time Doug had expressed an interest in Karen McKinney, the beautiful music major mentioned earlier, but I wasn't willing to give up on her. So we agreed to both take Karen to the Stravinsky concert. I don't remember which orchestra played that night in Beckman Auditorium, but it was a wonderful evening. They played the *Firebird Suite*. Widely considered one of the most important and influential composers of the 20th century, it was an honor to see Stravinsky conduct.

Doug's interest in Karen must have been stronger than I thought, since he brought up her name about 30 years later during a visit. But I was entering my no-dating-until-finished-with-graduate-school stage, so I probably should have given him the go-ahead.

Doug invited me to his parents' home in Washington State during spring break. They lived in Aberdeen, a paper mill and fishing town of about 15,000 people. Doug's father, William, was a prominent doctor, and they had a large house on a hill overlooking the town. It was a beautiful home, and he was particularly proud of his collection of classical music and his hi-fi sound system. Doug took me to visit a paper mill, which was quite interesting, with sawdust flying everywhere. I remember Doug's mother was Lutheran and very kind. His younger brother was smart and would also attend Caltech one day.

I especially remember the bedroom I slept in during the visit, Doug's normal bedroom. The bed was high and poofy, very comfortable. Under the bed was a large box of Marvel comic books: *Spiderman*, *The Incredible Hulk*, *X-Men*, *The Fantastic Four*, etc. This was long before all of them became blockbuster movies. Fascinated, I read as many as possible before going back to school.

Spring trimester, Steve Landy moved in with us. He brought a mattress that was plopped down in a corner of our dining room, where Steve slept each night—kind of Bohemian style. He was trying to learn ancient Greek at the time, a goal he continues today. He convinced Doug and me to include Greek quotes under our yearbook graduation pictures. Mine was to say, "Where is my parsley?" and Doug's would respond, "Here is your parsley." We submitted the quotes, using Greek letters, but apparently the publishing company couldn't handle the challenge, so our photo quotes in the final yearbook are blank.

My uncle Carl, aunt Ruby, cousin Deb, my parents, and sister Janis all traveled to California for my graduation. Carl was in the hotel management business then, so they stayed inexpensively in good hotels the entire drive to California. Janis was enrolled at Hamline University and was doing well despite all her medical issues. My father had worked two jobs to help pay for her education. I invited them all to our house on East Villa for a graduation banquet. We had to use the ovens both in our kitchen and our landlord's on the first floor to cook for everyone. It was a feast. Everyone got a deluxe Swanson's turkey TV dinner in his or her own metal tray, with newspapers spread over the dining room table, and a humongous watermelon was divided for dessert. They were all impressed.

Everything considered, my years at Caltech were perhaps the most interesting and important of my life. I met dozens of impressive people and learned a ton. Graduating with an overall GPA of 2.9 (B = 3.0) was better than my predicted GPA of 2.7. I was 26th among the 40 graduating physics majors, not good enough to merit a National Science Foundation fellowship, but not bad. On the Graduate Record Exam (GRE) in physics, I scored a 9.2 on a scale going up to 9.4. Nationwide, the 99th percentile for all those taking the physics test, and presumably trying to go to graduate school in physics, started at a score of 7.6. I, and most of my fellow students, were clearly the one in 100 that Caltech told us we were. Caltech made the rest of my career a lot easier.

Graduate School

It's interesting how the path of one's life depends on chance occurrences or quick decisions. My path to graduate school was probably set back in a high school class where I took a quiz that included questions about my future education plans. Knowing that scientists often study beyond a four-year degree, I jotted down that getting a PhD was my plan. Several nearby students laughed at me and one said, "No one from Johnson ever gets a PhD." That made me angry and resolute.

When my Caltech classmates started talking about graduate school, I needed a plan. Knowing that more than half of us would continue on, it seemed like the normal thing to apply to graduate schools. I was reading a book about low temperature physics at the time that included fascinating ideas and phenomenon: liquids that crawled up a beaker and over the edge, fluids and electrons that would circulate forever, and lots of unexplained science. So without much thought, I decided that low temperature physics would be fun. I had to put something on a form, and that's what popped out. My roommate Doug made a similar choice and may have influenced me on that decision.

I loved building things, so the only question was where the best low temperature experimentalists were. John Wheatley, at the University of Illinois in Champaign-Urbana, was one of the best. I applied to three schools—Stanford, the University of Illinois, and the University of Minnesota. Stanford turned me down, probably because I didn't have an NSF fellowship (only three or four from any college received them). I visited Professor William Zimmerman, who was doing superfluidity experiments at Minnesota, and I was later accepted there as well as Illinois. Both schools offered teaching and research assistantships worth about $3,000 per year. Illinois was my choice because of the famous John Wheatley, and perhaps I wasn't ready to go back to Minnesota yet.

Just after I accepted the assistantship at Illinois, Wheatley decided to move to San Diego and took his whole laboratory with him. So I spent a lot of my first year at Illinois building a refrigerator. One of Wheatley's students had been A. C. Anderson, a.k.a. Ansel, a.k.a. Andy, then a young professor at Illinois with a reputation as a friend of graduate students. After an interview, I chose him as my official research advisor. He was a great choice.

But even before arriving at the University of Illinois, one major decision had been made—I bought my first real car. I took out a five-year loan and bought a brand new turquoise Mercury Cougar. I liked the way the rear lights blinked sequentially. My father had to co-sign the loan since I had little credit history. The Cougar cost $3,000, my entire yearly salary; many people thought I was crazy. I have liked fast, cool cars my entire life, and 1967 was when it all began.

First Year at Illinois

Work at Illinois started right away during the summer. I got a dorm room in Daniels Hall, a multistory building right next to the physics building. A basement room in the Loomis Physics Building housed my desk and those of four other people. Time in the nearby low temperature lab was mostly learning how to solder copper pipes and brass valves for a new scientific refrigerator. Andy reprimanded me early on when he found me cleaning scorched test tubes. "You're costing me $5 per hour. Just throw out the old glassware." I learned that my time was more valuable than glass.

We erected a screened room to keep TV and radio radiation from heating up our experiments and two heavy brick pillars filled with sand supported our "floating" apparatus. Even cars driving by could cause slight vibrations that could heat up the super low temperature experiments we would be doing.

Before classes started in September, I moved to a room in a house on Green Avenue about a half mile from the physics buildings. My room had windows overlooking a hamburger joint, and I often sat by the windows on muggy days to catch a breeze and watch people eat in their cars below. I also signed up for a food program at a place called the Baptist Student Union, which might have been a mistake. They had many portly food servers who packed my tray with too much food. My weight went from 180 to about 205 pounds very quickly before I got it under control.

Across from my bedroom/study was a fellow grad student, Boyd Poston. Boyd studied in the finance department and taught some classes. Somehow he was paying for his entire graduate education with a single investment of about $3,000. Each year he invested that money in stocks chosen for certain financial ratios he calculated. When the value of his investments doubled, he could stay another year. Upon learning his approach, I invented a completely different method to make money with stocks, which I called the "harmonic oscillator method." After I experimented with my idea for a while, I challenged Boyd to see who could make the most money, on paper only. My method worked better than Boyd's while the market was going up—not so well when it fell. About forty years later, I tried my method for real, with good results, a story I'll tell much later in this book.

Classes started in September and didn't seem too difficult. One, with a textbook by J. Jackson, *Classical Electrodynamics*, seemed a little overly complicated in my opinion. But with my strong Caltech background, I kind of glided though all my graduate classes.

I signed up to play on the Daniels Hall flag football team that fall and we did very well. We had an extremely quick captain who was also our quarterback. I played offensive and defensive end, still using my strong legs to get by bigger people. We made it to the intramural championship game, where we faced an undefeated team from the law school. The law students all seemed to be 6 feet, 4 inches tall and very quick, and they threw a lot of high passes. We lost.

I also worked as a teaching assistant my first year at Illinois, and it was a hoot. I was one of about 10 graduate students assigned sections of the introductory course in physics for medical students. My group of 20 aspiring doctors and nurses all needed to pass physics to be considered for later medical school. Consequently, about two-thirds of the students asked for tutoring help at least once. I posted tutoring hours, but occasionally if I had worked too late in the lab or studying, they'd find me on an air mattress under my desk, napping. My sincere goal was for every single student to get an A or a B in the class, and I almost succeeded.

Every class session I taught began and ended with a five-minute quiz, with cartoon problems I had devised. For example: "Evel Knievel on a tricycle rides off a 200-foot cliff at five miles per hour. If a pond is five feet from the cliff, will he survive?" I knew that the only thing my class cared about was passing the big tests, so I gave them a lot of practice tests. Also, I kept them alert by occasionally throwing liquid air from a beaker onto the floor, which quickly spread throughout the entire room and created a cloud. Our low temperature lab had lots of liquid air available. There was no sleeping in my class.

Once, a nursing student saw her midterm test results near the end of a session and burst into tears. She ran out of the room and I wasn't sure what to do, but I went out to the hallway bench and comforted her. She had never received less than a B on any test in her life. I encouraged her, and later she got an A in the course.

I wasn't entirely successful—two of my students got Ds. One young woman often came in for tutoring. Her father had even offered to pay for a "real tutor." On the morning of the final exam, I worked with her for at least two hours. On one of my practice problems, I had a moon Dumbo circling the planet Jumbo and had her calculate the orbit time. During the actual test, one problem had a moon Tom circling a planet Jerry—almost the exact problem I had just briefed her on. From across the large lecture hall, I heard her say, "Oh, no, I should know this." Unfortunately, she didn't.

Clearly, I loved teaching and could very easily have spent my life in education. Nagging in my subconscious, though, was a concern that I needed to make money, good money, so I would never experience the struggles of my father.

In Anderson's lab we were starting to do actual research with our first refrigerator. A Polish scientist had published work showing an unusual electron conductivity result between 0.3 and 1 degree Kelvin, or about -459 degrees Fahrenheit. We showed that was incorrect and published the first evidence of an electron-electron scattering in platinum and silver. This turned out to be an important paper that was published in the most prestigious physics journal, *Physical Review Letters*. We even had theoretical support from John Bardeen, still the only physicist to have won two Nobel Prizes. I actually spent a couple of hours of one-on-one time with Bardeen discussing our work.

The Advanced Research Project Agency (ARPA) in Washington supported all our research. Even then, they were supposed to focus on science with defense implications; their name was later changed to DARPA. How measuring low temperature conductivity applied to defense was left for my advisor to explain. Andy's graduate students never got very involved with seeking or negotiating research funding.

Toward the end of my first year at Illinois our first super refrigerator was coming together. We needed to get below 0.3 degrees Kelvin to do really new research. This fridge had an outer thermos (a vacuum Dewar about five feet long and nine inches in diameter) filled with liquid nitrogen at 77 degrees Kelvin, a inner Dewar of liquid 4He at about 4 degrees Kelvin, a small chamber of pumped 4He at about 1 degree, and a final chamber of 3He at about 0.3 degrees. To get below that, we needed to construct something called a dilution refrigerator, which was very rare at this time.

The first dilution refrigerator had been built in 1964 in the Netherlands. To build one, I had to learn all sorts of new skills: sintering copper powder, building thermometers, checking for leaks, etc. Most importantly, I designed and built the entire plumbing layout for all the chambers mentioned above, all the valves, and connections to pumps. It was very complicated and required three-dimensional visualization and drawings. Ultimately, it all worked and we were able to cool things down to 0.015 degrees Kelvin, sort of the moving record at the time. Unfortunately, it was not quite cold enough, as will be explained later.

During the first year, I began to write to Steve Landy, telling him what a great school Illinois was. I hoped he might join me so we could continue our basketball and tennis competitions. Illinois is in the Big Ten Conference and the football and basketball teams at Illinois played in large, impressive buildings. Thousands of students would flow in river-like streams from all directions before big games. The teams weren't great, but it was a lot more exciting than "sports" at Caltech. I told Steve about all of it and my strategy worked. He accepted a research assistantship at Illinois for the following year.

As my first year at Illinois ended, I had passed all my classes and, sort of, automatically received a master's degree in physics. To begin a doctoral research program, I needed to pass a qualifying exam. The test caused panic among most students, but I looked at questions from earlier exams and they didn't seem too hard. I passed easily.

Illinois Year Two (1968–69)

The University of Illinois graduate school had a language requirement at the time, so I began a German class during the summer of 1968. I was acing the course during the first few weeks, riding on my two years of high school German. Then they started using new words. It's hard to understand how two years in high school could be covered in only three weeks, but it happened, and I started to struggle. Fortunately, the language requirement was reexamined by the Physics Department and was simply dropped about that time. Within a few days I was relieved to be out of the class. Nevertheless, many German words are lurking deep in my mind still, and pop out occasionally.

Steve arrived in the fall and we found a place to stay about ten blocks from the physics building on Hill Street. An elderly black couple owned the furnished two-bedroom home and only charged $105 for rent. Steve and I shared a bedroom, and a huge guy from Chicago took the other room. He was a very aggressive and surly roommate. Once I got in a fight with him over an easy chair in the living room that faced the TV. I sat in his precious chair when he was absent and he tried to throw me out of it when he came into the room. We ended up breaking a lamp after wrestling around the room.

The biggest change in life at Hill Street was that we needed to buy and cook our own food for the first time. My dinners soon fell into a pattern: one day, TV dinner; next day, steak; and the third day a giant salad (usually a half-head of iceberg lettuce). By the end of the week it was probably a balanced diet. Another big change was washing dishes. I seemed to be the first to clean up when I couldn't stand the piles of dirty dishes any longer—the curse of being neat. Also, Steve and I started saving glass soda (pop) bottles. We left the six-pack empties in their cartons and piled them along a kitchen wall. The stack increased gradually, and every six months or so, we would take them for refunds at the market. One time we filled my entire Cougar, seats and trunk, with empty bottles, maybe a thousand. And we lucked out. The price of an empty had just increased from 3 cents to 5 cents. We made about $50.

In addition to the three humans, two other species lived at 1103 Hill Street. One was Fritz, a large tomcat who was quite paranoid. He must have had an earlier owner who held him a lot because he needed a lot of affection and seemed nervous all the time. Every night he went out and got beaten up by the other cats, often having scratches on his face. A few years later the first feature length cartoon came out, surprisingly with the title *Fritz the Cat*. It had great jazzy music, I bought the record, but the movie was X-rated.

The other species living with us was cockroaches. Going into the bathroom during the night and turning on the light was a horror. Dozens of critters would scatter in all directions. We sprayed around the bathroom walls once to get rid of them, but the creepiness and cockroaches remained.

Steve and I often played pickup basketball on the courts between Daniels Hall and the Loomis Physics building. We played well together, especially the pick-and-roll. There always seemed to be games going on. One time one of Steve's professors, Lorella Jones, walked by and gave Steve a dirty look. She apparently didn't understand the intrinsic value of basketball.

A lot of interesting things happened in the lab that year. We published several papers in scholarly journals, with titles like: "Thermal Conductivity of Mylar between 0.05 and 0.5 K." Useful, but obscure stuff. I started receiving requests for paper reprints from places like Kyrgyzstan and Jordan.

For fun we poured liquid nitrogen on the floor near closed lab doors, which skittered under the bottom crack into the hallway just as women with open-toed shoes walked by. We could hear them yelp through the door. Our lab group all still had a confused understanding of females then.

Once, when I flicked on a light switch, a wrench flew across the room and stuck to a giant magnet. Scary! We couldn't figure out how turning the switch made the wrench fly and concluded it was just coincidence.

We once spilled a lot of liquid mercury on the floor that scooted all over. Mercury wasn't recognized as a bad agent at the time, so we just tracked it all down and nudged it back into a bottle. If that happened now, the entire building would probably be evacuated and a hazmat crew brought in to clean it up.

My most memorable day was in the nearby machine shop. I was milling electrical resistors to make very thin slabs of carbon, about one thousandth of an inch thick. They served as resistance thermometers for our experiments, sensing to very low temperatures. Milling them was difficult and slow work because they were so fragile. After working from 10 a.m. until about 7 p.m., I had produced only a dozen of the tiny devices. Everyone in the machine shop had gone home when I finally got to the door and turned off the lights. I had placed a sheet of paper with all the sensors on a shelf near the door. When I opened the door, the breeze came through and lifted the paper, throwing all my sensors to the floor. I turned the lights on to see that every single one had been broken. Luckily, no one was around to see me cry.

Illinois Year Three (1969–70)

In 1969, a person with a familiar sounding name, Miriam Cox, called to ask if I would be willing to give her rides to Minnesota during University of Illinois breaks. She had seen my name and number on a bulletin board as having a car and that I was looking for riders to Minnesota. Miriam had attended Johnson High School and remembered that I had officially inducted her into the National Honor Society when she was a junior. She said that she and both her parents remembered me clearly as the one who nearly set the stage on fire with my chemistry demonstration during a program one freezing January night. Miriam was supposed to read a poem she had written that night but had laryngitis, so her teacher read the poem while she sat on a stool on the stage.

We had lots to talk about on our drives. While I was studying at Caltech, Miriam had graduated from Augsburg College in Minneapolis with a major in Spanish, and then received a Mexican government fellowship to study at the National Autonomous University of Mexico (UNAM). UNAM, located in Mexico City, was the top university in Mexico. She had found a family to live with there and was ready to start classes only a short time before the Summer Olympics were scheduled to begin. This was a time of great unrest, and over 50,000 students had marched through the city to protest regressive government policies. A short time later, at least 300 students and others were killed at another protest, which led to the Mexican army occupying UNAM and shutting down classes. There were actually tanks scattered around the campus.

Fortunately, Miriam was asked by the U.S. Embassy to volunteer as a guide at the U.S. exhibit at the Olympics. She also was able to get tickets to four Olympic events, including the track and field event where Bob Beamon set the record of 29 feet, 2 inches in the long jump, which stood for over 23 years. She was also present for the famous black power salute of Tommie Smith and John Carlos during their award ceremony. Meanwhile, her classes didn't actually begin until the second semester. Thankfully, the Mexican government continued to pay her $100-per-month stipend.

Miriam and I shared the drive between Illinois and Minnesota during winter, spring, and summer breaks for one year. I had previously been alone during the boring eight-hour journeys, so having someone from my old high school to talk to was a great improvement.

Once, driving alone, I had gotten a speeding ticket in Wisconsin, and my father had to deposit money by telegram before I could be released from the police station (they didn't allow credit cards and I didn't have enough cash). Another time I nearly fell asleep on the road and pulled over on a farm road for what turned out to be a six-hour nap. I needed a co-driver.

The long drive with Miriam was very pleasant, and heading into the sunset as we came home vaguely romantic. She was an enjoyable companion and I felt very relaxed sharing driving with her. Also, her mother was sweet and treated me kindly at her house after our long trips. I began to think of Miriam as a serious romantic possibility—after I finished graduate school, of course.

Once Steve and I invited her over for her birthday and made a cake. I really liked to look at her legs. She received her MA degree in June of 1970 and returned to teach in Minnesota. To maintain contact, for two years I called her around New Year's, just before returning to school, to ask if she was married yet, but too late to ask for a date. It indicated my interest, I figured. She thought I was nuts.

In the winter of 1969, Steve Landy looked around my basement office and invited all the students there to join us on a new graduate school intramural basketball team. Most of the students were Asian and rather short. We soon had enough to form a team. We named our team the Dragons. Some team members could actually shoot a basketball, long set shots, but none of us were very good.

We lost a few games but were competitive. Then we met the Umonga, an all-black team. Every player seemed to be at least 6 feet, 6 inches tall, except for one. I remember looking to the other end of the court during warm up and saying, "At least one of them is short." When we came together at midcourt we could see the very short guy's T-shirt—it said NEW JERSEY ALL PRO.

They were ahead 20 to 0 when I was fouled. Lining up my free throw, I remember thinking, *These might be our only points.* I missed both shots.

Later, they stole the ball and went on a fast break. No one from our team even bothered to chase them down the court. Soon all five of the Umonga were at their basket, passing the ball around. One guy started bouncing the ball off the backboard to shame us. As soon as I got down to defend, he simply tapped it into the basket. I think the final score was about 95 to 10.

Living on Hill Street became much more pleasant after the rude Chicago guy left. His room was taken by Terry Follinsbee, another student of A. C. Anderson and a Canadian. We all agreed that our landlords had been charging too little for our rent. So we went over to visit them and said we would be raising our rent to $135 per week, or $45 for each of us living at their property. The woman explained that she had worked at the university and loved the students, so they hadn't even thought about increasing the rent.

Later that winter, in 1970, I received terrible news from back home. My grandparents, Hans and Ida, had died in a horrible fire along with my teenage cousin Dennis who was living with them at the time. Ida was 85 and Hans was 81 when this happened. Hans had a bad habit of smoking cigars and apparently one fell from the living room couch ashtray into a pile of newspapers, starting a fire. He went to the basement to get a hose, but the hose became kinked when he reached the living room. He tried using buckets from the kitchen but was overcome by smoke. Ida and Dennis died in the upstairs bedrooms.

I drove home from Illinois and walked through the burned home with other family members. I am still haunted by the wet, sweet, sooty smell and black awfulness of what I saw. A shadow image of Dennis's body was still clean on the sheets with ash all around. In the dining room, filled with so many memories, all the photos and artwork on the wall shelves were gone. The wood pillars I had climbed as a young boy were charred at the tops and oddly unaffected near the bottoms. The cabinet with their precious salt and pepper shakers had been destroyed and a pile of ceramic pieces were mixed with ashes on the floor.

Until then, at age 25, death had been an abstract concept to me. But now the extended family had lost its center post and it really hurt. I stayed through the funeral and mourned with the family before heading back to school.

At the university, and around the country, 1969 and 1970 were years of great unrest. In February, Illinois students protested the installation of Illiac IV, one of the fastest computers in the world, when it was discovered that two-thirds of its time was reserved for the Department of Defense. Soon after, the ROTC lounge of the university armory was firebombed. Hundreds of students continued to protest, and in March a curfew was imposed and enforced by the National Guard. I wasn't very political at the time and certainly didn't want to waste time protesting, so I adjusted by simply living in the laboratory. Sleeping in the screened room on a leaky air mattress became normal. I let liquid air bleed through a needle valve into the leaky air mattress, so I wouldn't sink to the floor by morning. That also kept the mattress very cool in the muggy summer months.

Those were also truly exciting days for the space program. Neil Armstrong, on Apollo 11, walked on the moon on July 20, 1969, while Apollo 13's near disaster was in April 1970. One moon mission coincided perfectly with one of my experiments. As the experiment cooled down, a three-day process, the astronauts traveled to the moon. They then circled the moon for one day while I did my experiments.

Then the refrigerator was warmed up, taking another three days, while the astronauts came back to earth. All the way through this I was listening to a non-stop broadcast about the mission on a radio outside our screened room. I felt like I lived the space missions with the actual explorers.

Between experiments, a lot a time was spent building devices and parts for our measurements. I remember polishing a lot of beryllium (Be), a very unusual metal. Beryllium has unique properties: extremely light, strong, brittle, high thermal conductivity, and high sound velocity, perfect for some of my experiments. Unfortunately, its dust is also highly toxic and causes cancer (berylliosis). I polished Be in a vat of oil to prevent particles from becoming airborne. That didn't stop fellow grad students from taping articles on lab walls about people who had died from Be poisoning.

During the summer of 1970, someone set off a massive bomb at the University of Wisconsin physics building at 3 a.m. The bombers were protesting the Army Mathematics Research Center, which was actually undamaged. However, one basement low temperature researcher, working late, was killed and three others injured. They were doing superconductor research, and A. C. Anderson knew their professor very well. We were all nervous and kept alert while working late after that.

Illinois Year Four (1970–71)

The summer of my fourth year at Illinois was almost entirely wasted. Our super refrigerator developed a leak that prevented any experiments. We used a leak detector to check all the lines and chambers of our multi-stage refrigerator but couldn't find the source of the leak. We had the worst possible problem, a leak that only appeared after cooling to low temperatures, a three-day process. Eventually we found the leak by hanging our Dewars partly open and using long needle probes and mirrors to see where the leak was located, but it took three months!

I had started work on my thesis research by this time, which was focused on the mysterious phenomenon known as Kapitza resistance. Some of the greatest Russian scientists had worked on the problem—Khalatnikov, Landau, and Kapitza himself (who later led the Russian Academy of Science and won a Nobel Prize). The problem was that experimental measurements and theoretical predictions differed by as much as 1,000 to 1, and no one knew exactly why.

All my research had delved primarily into thermal conductivity, how heat flows at very low temperatures. Why is this important? One reason is that at very low temperatures, materials and science itself are very clear and fundamental. Without heat to cause vibrations, atoms settle into their most comfortable positions, their lowest energy states. This is why 4He becomes a superfluid, one that will circle forever in a glass. It is in its lowest state and the entire glassful is essentially one thing, sort of like one atom. It can't stop circulating because it is "all together." The same general idea applies to super conductivity and laser beams, both very useful and valuable properties.

Knowing how heat flows is, therefore, key to experiments and understanding the results. It is also a key to any practical applications. Having to cool things down so far is, of course, a nuisance, which is why people have searched for "high temperature superconductors." Phenomena at low temperatures are freaky, and it would be really nice if they happened at higher temperatures. For example, a room-temperature superconductor would make quiet, smooth, super-fast maglev trains practical, and possibly even floating cars.

Between experiments, my work involved making large chunks of epoxy to machine into special shapes. Increasingly, it also involved polishing materials, especially metals. Using all my telescope experience, I polished about ten metals as carefully as possible, using abrasives and then chemical and electrochemical techniques to remove the final atom layers from samples so there would be no stress, scratches, or other disturbances on the metal surfaces. This was critical to the value of the experimental results.

In the actual experiments, helium liquid and other materials were placed on these polished surfaces and the thermal resistance across the boundary was measured. As the results came in, I did calculations on a Wang calculator. The Wang had orange nixie-tube readouts and could store several constants, and was capable of doing logarithms and other functions for the first time. The old slide rule was slowly being replaced.

I agonized over the first few results and devised elaborate theories to explain the results. It seemed I could explain almost anything given enough time, sometimes with ludicrous ideas. Creativity has never been my problem.

Professor Anderson gave great advice at that time saying, "Take more data." He was right. As more results with more metals came in, what was really happening with the Kapitza resistance was starting to become clear.

Then the work slowed down again. In December 1969 the country held a draft lottery to determine who would be going to war in Vietnam. I received a fairly high number, in the top third, and had to go for a physical to see if I was qualified. We all had to strip off our clothes and put them in lockers in the Minneapolis Armory and follow a yellow line on the floor for a series of tests. We were weighed, poked, and tested and finally sent in to talk to a doctor/counselor. That's when I remembered that I had forgotten my written doctor's excuse in the locker with my clothes. I quickly scrambled back to get the signed paper, cutting across many yellow lines. Soon army people were screaming at me. I thought I might be shot. But I managed to get my paper and hand it to the recruiting doctor. It said I had asthma and other ailments and should be excused. The doctor looked at the paper and said it was a borderline case and that I could volunteer if I wanted. Although my father served in WWII, he never described it as a wonderful experience, and my mother had always worried about the dangers of military life. She didn't want people shooting at her little boy. So I told the doctor I wouldn't be volunteering.

One of my fellow Illinois grad students, Steve Smith, wasn't able to avoid the draft, so Professor Anderson said he should have priority to finish his thesis work before joining the Army. We fellow researchers gave Smith complete access to the super refrigerator. He did finish up in time to get his doctoral work done before going to war.

During this time, we started building a second screened room and a new refrigerator. It was essentially a duplicate of the first one but would speed up the experimentation of Anderson's students. There were about five of us by then.

Overall, we were publishing scientific papers regularly that year, and I started looking for someone to type my future thesis. In those days they had to be typed perfectly, according to precise rules, and were best finished by professionals.

Illinois Year Five (1971–72)

Okay, I'm going to try to explain my doctoral thesis research. Little vibrations called phonons are the heat carriers at low temperature. They are similar to photons, which are particles of light. When a phonon hits the boundary between two materials, its path is bent, like a lens or prism for light. In the denser material, the paths are bent toward the normal to the surface. The amount of bending depends on the velocity of sound in the materials, which for a solid-to-helium boundary can be a factor of 20 different, i.e. it bends a lot. Many phonons in the denser material hit outside of a critical angle and are totally reflected. In general, the phonons are struggling to get through to the other side. Another way to describe it, which would make sense to electrical people, is to say there is a great "impedance mismatch" causing reflection.

But experimentally, a lot more heat gets through than would be expected from the above discussion, about 100 to 1,000 times more. Something was way off about Kapitza's theory. It turned out that the phonons were penetrating a little into the other material, sort of running along the interface. They weren't totally reflected right away. It's a little like earthquake waves: some run along the surface while others go through the bulk of the earth. That's how one can tell how far away an earthquake occurs. Anyway, while phonons in my experiments ran along the surface, they were scattered or absorbed into the other material. I solved this by including a new factor in the usual equations for what happens at an interface by including a complex number that accounted for the absorption or scattering. Then, in various ways, we determined the absorption either by including electron effects or estimating physical scattering. The method worked for numerous metals and helium combinations, correctly predicting the high thermal conductivity that everyone had been seeing. The work took about one year. Many improvements have been made since then to improve the model, for example by including quantum effects, but generally my new theory was correct. I had solved a problem that had been a mystery for ten or twenty years.

I typed a first draft of my doctoral thesis, making an average of 50 typing mistakes per page. Using a lot of white correction fluid, the mistakes were fixed and then turned over to a professional typist. When you held up a page of my draft with light behind it, it was pockmarked with all the corrections—very funny looking. But the job got done and A. C. Anderson and I published a journal version in *Physical Review B*.

By June 1972 I was essentially done with one big hurdle remaining, the oral doctoral exam. In addition to Andy, three other professors were on my advisory committee, including John Bardeen. In December that year he would win his second Nobel Prize for collaborating on the first explanation of superconductivity, known now as the BCS theory (for Bardeen, Cooper, and Schrieffer). Bardeen's other Nobel Prize was for the co-invention of the transistor.

Giving a talk in front of those four professors was not easy for me. In addition to explaining my thesis work, I had to be able to answer any physics questions that might come up. I was very nervous and can't remember a single thing about the grilling that day. But Andy did tell me what Bardeen said after the exam: "Terrible oral defense, but a great thesis. He passes." I'll never know how close to failure I came that day.

In addition to the physics, Steve and I were doing a lot of serious bike riding. Back in my first year at the University of Illinois, I'd purchased the only good bicycle I've ever owned. The frame had the words WORLD'S GREATEST BICYCLE BY SEKINE painted on. I'm not sure if it was the world's greatest bicycle or just the best one that Sekine made, but it cost about $200, a fortune at the time.

By my fifth year at Illinois, long-distance biking had become a frequent activity. Steve and I rode most weekends, often with Terry Follensbee and another physics grad student, Mary Alice, to state parks all over Illinois. We'd camp out overnight, with me sleeping in a plastic tube tent hung between trees. Illinois is great for bikes because it is so flat and it has rarely used country roads everywhere between the actual highways.

Riding for several hours to a different park was fun, and it got us ready for an upcoming bike adventure. Steve and I had planned a major trek from Illinois to Minnesota along the Mississippi. I had tried to scope out the path after dropping Miriam off in St. Paul once. With no good bike maps at the time, my ride back to campus took 14 hours on small roads while I checked to be sure they were safe, and I had only barely made it to Illinois. So I rested and then drove freeways back to school. There would be some unchecked sections along the path.

On the first day of the big trek, Mary Alice drove us to the Mississippi River, dropping us off near Davenport, Iowa. Terry, Steve, and I started off on what turned out to be a 600-mile ride that lasted about one week.

After only a few hours, I caught my bike tire on a road edge and went down. Scratches on my legs and hands weren't the worst things; my front tire had a 20-degree bend. Using brute force to straighten the tire as much as possible, we continued on with my tire going woomp-woomp-woomp the whole way. I had to disable the front squeeze brake, so I couldn't stop well, but we made it to Dubuque and found a bike shop. After a long wait, I got a new front wheel and we were off again.

The trip was difficult. The temperature lingered in the 90s most days and the bluffs along the Mississippi went up and down about 600 feet every time a stream flowed into the main river. It would take an hour to go up some of the hills and then only five minutes to go down the other side. Terry was frightened by the descents at 50 miles per hour, so he often lagged behind. I used my hand brake so continuously on one hill that a tire overheated and melted the glue holding the sew-up tire to the wheel frame. The tire rotated to stitch-side-out and blew at high speeds, but we had lots of tools to make repairs. After that, Terry considered hitching a ride in a pickup truck back to school, but no pickup came for a long time and we convinced him to continue.

Another day we only rode for about two hours, 30 miles. We climbed a very steep hill, passed farms at the top of the bluff in stifling heat, before coming to a dangerous plunge down to a river town. Flying downhill, we rode to the main drag and started looking for a motel and a place to eat. We'd been camping out until then and were totally exhausted, so showering and a real bed seemed overdue. For supper, each of us ordered two steaks and two malts.

The final day's ride lasted over 100 miles along the river and then through the Twin City suburbs to my parents' house. It was near 100 degrees the whole way and we all flopped down in the backyard. My mom and dad came out with ice water and later hosed us all off before letting us into the house. Everything considered, our trek gave us a new viewpoint of life along the Mississippi and of ourselves.

Although I was effectively finished with my doctoral work, Andy Anderson convinced me to hang around a bit longer to publish more journal papers. But then it was time to start lining up a future job. One memorable August week at the 13th International Conference on Low Temperature Physics in Boulder, Colorado, I saw my old roommate, Doug Osheroff, again.

Talking with all the luminaries of the low temperature world was very exciting for me, and many wanted to understand the results of our work on Kapitza resistance. But the big shock for me was a special session held to discuss a new discovery by Osheroff and his colleagues at Cornell. They reported the discovery of a new transition in 3He to what seemed to be a superfluid state. This was something everyone had thought possible, and all the top labs were competing to find it. Doug used something called a Pomeranchuk cell to get to even lower temperatures than a dilution refrigerator. At somewhere around 2.6 mili-Kelvins, he saw some unusual blips in his experimental chamber, which signaled the transition.

A little physics needs to be added here to explain why all this was important. There are two types of fundamental particles: fermions with spins of 1/2, 3/2, 5/2 . . . , and bosons with spins of 0, 1, 2, 3 . . . . Why two types of particles exist, I can't easily explain, but they behave very differently. Bosons like to hang out together, as mentioned, which explains why 4He is superfluid and why photons can form coherent laser beams. Fermions seem to hate one another and stay apart. That's why superconductivity, which involves electrons (fermions), was so hard to explain. Bardeen and his colleagues showed that the electrons form funny pairs, now called Cooper pairs, at low temperatures and no longer act like fermions. That's why they won the Nobel Prize. Same thing for 3He, it's a fermion so superfluidity should be impossible. Doug and his Cornell co-researchers found that 3He also can form pairs at sufficiently low temperatures.

At the conference, I saw that Doug was his old self, wearing non-matching loud socks and enjoying the notoriety. But I learned two more important things that week. One, I should have had an even better refrigerator (I missed the same discovery by only 14 mili-degrees). And second, and more importantly, if you want to discover something in science or anything else, it's better to look where people haven't already been. My solution to the

Kapitza problem was nice, but many had dug there already. Doug's discovery opened up whole new areas of science, research, and understanding. I remain jealous to this day.

The first 27 years of my life, my Jumping Years, included achievements far beyond the expectations of my relatives or Johnson classmates. I had experimented, tested myself, taken risks, and learned as much as I could.

I had developed a hero list. The list included Galileo, Newton, Einstein, Bardeen, and Dick Fosbury (physics student and creator of the high-jump flop named after him). I would later add Steve Jobs and Elon Musk to the list. They all were or are inventors and people who changed the world forever, often quietly and surrounded by skeptics. That's what I wanted to do. It was time to get to work.

Maternal Grandparents Ida and Hans around 1916

Ron's mother Fern, 1916   Fern's H.S. graduation photo, 1930

Ed's father Barney, 1953

Ed's H.S. graduation photo, 1936

Fern and Eddie courting around April 1942

Eddie and Fern's wedding, September 7, 1943

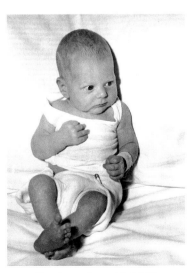

Ron born June 18, 1945

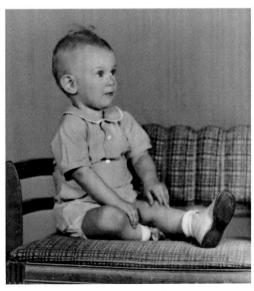

One-year birthday

Family portrait August 10, 1949
Janis age 2, Ron age 4

Ron and Cousin Marsha at Aunt Myrtle's wedding, 1951

Sunday dinner at Grandma and Grandpa's.
From left: Ed, Ron (reaching), Fern, Hans, Ida, Janis, Dennis, and Myrtle.

A recent photo of our house at 950 Burr Street after being painted blue.
We lived upstairs with Uncle Ernie's family downstairs.
Note the upper porch with many windows,
a perfect place for a telescope observatory with windows removed.

Ron as school police captain in 1955 at Ericsson School (a great honor). Lieutenant Kenny Ryberg in back.

Ron climbing and jumping, April 1956

Ron with cousins Dennis, Marsha, and in front Deb and sister Janis in 1959.

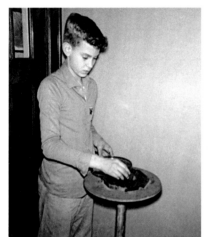

Building Telescope Number 2, winter to fall 1960

BUILDING his own telescope was Ronald Peterson who demonstrated the optics of telescopes for a blue ribbon at the Twin Cities regional science fair.—Staff Photo.

At Regional (blue ribbon) and State (red ribbon) Science Fairs.

PETERSON, RONALD E.
H.R. Rep. 3; Student Council 3; Library Club 3; German Club 2,3; Radio Club 3; National Honor Society 2,3; A Basketball 3; B Basketball 1; Track 1,2,3; Intramural Basketball 2; Cross Country 1

COX, MIRIAM C.
Honor Guard 2; H.R. Rep. 1; A Choir 2,3; Operetta 2,3; Class Rep. 1; Pep Club 1,2,3; Quill and Scroll 3; Spanish Club 1,2,3; Historical Soc. 2,3; Courier Staff 3; Governor Staff 3; Natl. Honor soc. 2,3; G.A.A. 1,2,3

High school graduation photos
Ron 1963, Miriam 1964

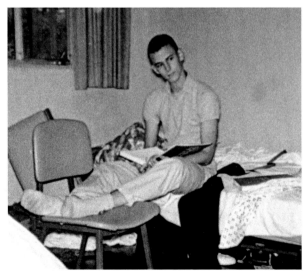

Caltech, 1963, Ron hard at work freshman year.

JUNIORS—FRONT ROW: T. Buckholtz, M. Beeson, J. Manke, B. Piccioni, S. Abrahamson. SECOND ROW: D. Osheroff, D. Balanis, E. Lippa, A. MacKay, G. Jennings, R. Peterson, G. Berman.

Lloyd House Class of 1967

Ron sets an interhouse record by clearing 5 feet, 11 inches.

Ron and Doug Osheroff graduate.

Grandma Ida shortly
before the fire.

Aunt Ruby and Cousin Jim by
salt and peppershakers

Janis's graduation from
Hamline, 1970

Miriam's graduation from
Augsburg, 1968

Miriam's family, 1968: George, Helen, Miriam, Ken, and Bill

Hosed down after our 600 mile bike ride
Terry, Ed, Ron, and Steve

Ron's PhD Advisor, Dr. A. C. (Andy) Anderson.

Some of Ron's refrigerator plumbing, good to 0.015 degrees Kelvin.

PART II

Working Years

inally out of school—the dam burst. All of my years of delayed gratification were ending. I could earn money. I could talk to women. I could go home.

For at least nine years I hadn't spoken to any child under 15 or even seen a baby up close. I had rarely been inside a church or a department store. I had 100 pairs of generally mismatched socks.

Getting a doctorate could often take seven or eight years and I didn't want to wait that long. So I had focused. The ability to focus and shut out distractions is perhaps my greatest strength and greatest flaw. But the years of sleeping in a basement lab had ended and I could find out how normal people lived. I was free.

Getting a Job

Search for permanent work began in the spring of 1972 while still at Illinois. Companies came to the university to recruit students, and I signed up to talk to General Motors and Honeywell. The General Motors recruiter convinced me I wouldn't want to work at General Motors. He kept talking about why advertising was more important than research and how we needed to come up with multi-billion-dollar ideas or the scientific work was worthless. So despite some great facilities and capabilities, GM was crossed off the list.

I sent applications out to dozens of places and received a job offer from Los Alamos in New Mexico. They were scaling up to do controlled nuclear fusion work using giant super-cooled magnets. My mother's words about returning to Minnesota weighed on me, and I also knew a couple of nice girls in Minnesota, including Miriam, so New Mexico was crossed off the list. A totally unsolicited offer as a post-doc also arrived from a small university in Hoboken, New Jersey, which never even got near my list.

Back in Minnesota around June, I visited Hamline University, where a long-time physics professor had retired. Speaking with the other physics professor there, who was the official "decider," was interesting. He said he had over 500 applicants for the position and I was unacceptable because I graduated east of the Mississippi River. He needed ways to pare down his list to reduce the workload and the last hire in the science department had been from east of the Mississippi. They also needed someone to teach astronomy and that wasn't on my resume. I left thinking, *What a stupid way to hire someone.*

I also visited the central laboratory of the 3M Company, where I really wanted to work. An interviewer said there would be no physics people hired that year. Offering to cut their grass didn't help. He said they never hire PhDs to cut grass, laughing.

In 1972 almost all government labs, companies, and universities still had hiring freezes due to uncertainties in the economy. With the Vietnam War, price controls, and political turmoil, everyone was in a wait-and-see posture.

Honeywell and 3M were the two biggest technical companies in Minnesota at the time. So Honeywell was the other obvious place to try since they had excellent labs and a good reputation. Although I had applied for a Honeywell interview at the University of Illinois, I got to the job office late one morning and my name was too far down the list to get an interview. So, back in Minnesota that summer, I called to see if the opening had been filled and learned it hadn't. They were looking for an expert in lasers and electro-optics, and I would have been tossed out during the university recruiting tour processes. They certainly didn't need an expert in low temperature physics. Over 15 colleges had been visited, I was told, and about 200 people interviewed. Despite all that, I was invited in on a Friday for a chat.

The research section chief who would do the hiring was not in that day, so I was allowed to talk to most of the scientists in the group who gave me a good idea about what they were looking for. The next day the section chief, Jaan Jurrison, would be back in town and a meeting was set up for Saturday morning. I spent the evening trying to figure out how to show them I could learn and do anything they needed.

Jaan Jurrison turned out to be a friendly man with an odd Estonian accent. I knew they wanted someone who knew about polishing mirrors, so I talked about building my high school telescopes and polishing many different metals in my low temperature research. Luckily, Jurrison had great curiosity about low temperature physics, superconductivity in particular, so he actually got excited about my thesis work. The Caltech degree helped as well.

Finally, a lucky coincidence boosted my odds of landing the job. During our talk, Jurrison got a phone call from Dr. Kalafalla, who was looking for his wife Aida, one of Honeywell's scientists. Jurrison mentioned he was interviewing a Ron Peterson. Dr. Kalafalla, my summer employer back at the Bureau of Mines, gave me an immediate, unsolicited endorsement. On our walk to the door, Jaan Jurrison said that he thought I was a good fit for the job. On Monday I had an official job offer from Honeywell over the phone.

I consider this whole thing rather miraculous. Missing the Illinois sign up sheet, having extra time to prepare for the Jurrison interview, and getting an on-the-spot recommendation from my past employer were all needed to get the job. I accepted the offer with a start date of December 4, 1972.

Getting Paid and Scrambling

My salary at Honeywell was set to be $16,000 per year, up considerably from the $3,000 I made as a researcher at Illinois. The consumer price index has risen 5.7 times in today's dollars, so my 1972 salary was equivalent to $91,000 today. In November, I moved home to my parents' house and told them I would pay $100 per month rent. They were happy to see me again, and I got room, board, and laundry service. It was a great time—I felt extremely rich.

Somehow I must have really impressed the Honeywell interviewers or, alternatively, they might have been desperate. My official job title was Principal Research Scientist. This was up one notch from where most people with doctorates were brought in. The group I joined was very diverse and interesting: Henry Mar (a Chinese ceramics engineer), Wolf Doerfler (a German

chemist), Hans Mocker (a laser expert), Aida Kalafalla (an Eqyptian bio-
physicist), Jim Ramsey (a mechanical engineer and part-time University of
Minnesota professor), Roger Heinisch (a physicist and imposing former ma-
rine), and Mark Gurney (another physicist and Caltech graduate). All had
PhDs. The entire organization I joined was called the Systems and Research
Center (S&RC), which had about 150 people. We shared space with the larger
Honeywell Avionics group in a facility called the Space Center, essentially
a big warehouse with labs and small cubicles for each scientist or engineer.

My first act after orientation was to turn in my moving expense report.
Henry Mar said it looked ridiculous. My total expense bill was 50 cents. I
had piled all my worldly possessions in and on my Cougar and drove straight
home from Illinois. I figured that Honeywell shouldn't be charged for that
because I was going home anyway. The 50-cent charge was for a large hunk
of cheese I had bought in Wisconsin. Henry said I couldn't ask for less than
$50 compensation, so I eventually figured out how to bump it up.

My first days on the job were spent reading about Honeywell and about
lasers and electro-optics, which was supposed to be my area of expertise. In
the library, I discovered that the word "laser" did not have a "z" in it. That is
how little I knew about my job.

I soon discovered that I'd need to learn even more about how to talk
to people, even socialize, things I'd never really learned. The first week,
as I walked down a hallway, a man I'd never met said, "Hello, Ron. I've
heard good things." I soon learned that the man was Matt Sutton, the Vice
President of the entire research lab. How did he know me?

Also, a tall spooky man walked by my cubical several times during my first
few days, crouching low, staring at me and grunting a little before moving on.
Eventually he stepped in and told me about filling out my time card and how
much he hated to do his. Ozzie Severson was his name, a quirky, delight-
ful Norwegian engineer with a deep voice, who later became my good friend.

Another memory—at the group Christmas party in mid-December, I
was forced to dance with Aida Kalafalla, the voluptuous biophysicist from
Egypt. The only thing that frightened me more than women was dancing
with them. There was also drinking involved. The world of work was turning
out to be quite different from my quiet lab at Illinois. I had entered a world
where lots of strange and interesting people lived.

Much of the work at the S&RC was government funded and supposed to support the aerospace parts of Honeywell. Honeywell had another lab that did commercial research. It soon became clear why they hired me—they were totally swamped with work. They threw me on three or four projects in the first weeks. One required me to write a final report on a project I had not participated in, using the notes others had taken.

Down in the lab, the smell of hydrogen sulfide was strong near the working exhaust hoods. The gas was used to make lead sulfide detectors that went on the Pioneer and Voyager space missions. The sensors detected the sun and the star Canopus to orient satellites. Some of our detectors have traveled over 12 billion kilometers and still work.

One of my first projects was to make a solar black coating on a 6-inch diameter sphere. A lab technician, Bob Daggett, was assigned to help me and we built a vacuum chamber fixture to rotate the sphere around two axes while multi-layer coatings were evaporated onto the sphere. The plan was to put the coated sphere at the center of a large parabolic reflector on a satellite to gather solar energy and, with a special sterling-cycle engine, produce electrical energy. Our method worked and I was given the assignment to carry the black-coated sphere to Wright-Patterson Air Force Base and deliver it to our customer. Well, a 6-inch black sphere looks exactly like the kind of bombs you see in cartoons, so the airport screener wouldn't let me get to the gate. I had to go back to the company to get an official explanation letter before I could get on the plane.

Another project involved designing reflective coatings for high-energy lasers, probably for future weapon experiments. This work really impressed my co-workers. Roger Heinisch, already a superstar in the lab and future Honeywell Vice President, had been trying to predict the performance of multi-layer coatings, which required inclusion of a small absorption of the coating materials. We had to know if the coatings would burn up when hit with powerful lasers. Roger had tried to model the numerous reflections and absorptions in ten or more layers, which led to incredibly messy equations. I recognized it quickly as being analogous to my graduate work. We could use the simple equations for normal reflections on a multilayer, non-absorbing coating, but account for the absorption by including an imaginary component for the refractive indices. I wrote new equations, Bob Daggett wrote the code, and we sent our computer cards off to a big computer. It worked, and we delivered a simple new program in a final report.

Things were going pretty well at Honeywell. Work on solar energy materials and devices had picked up, which would later spin off as a whole new research center. And, despite all the day work, I was also developing an actual night life.

Dating and Beyond

As soon as I returned to Minnesota I tried to date two girls. One, Diane, was a friend of my cousin Deb Holmgren. She was pretty, pleasant, and rich. Her dad owned a significant local business and drove a Pantera. A De Tomaso Pantera is roughly equivalent to a Ferrari, so I was pretty impressed.

The other woman was Miriam Cox, my Illinois-to-Minnesota ride-share companion. She was teaching Spanish at Harding High School. On all of my previous calls, asking if she was married yet, she had said no. But now when I called, she said she was seeing someone named Eric, a University of Minnesota exchange student from Norway. It was quite a blow. She did agree to attend my big graduation party celebration with all my relatives and Steve Landy, who had driven in from Illinois. I asked her if I could talk to her after the party at her place, but she said Eric was waiting there for her.

Since Miriam seemed to be unavailable, Diane became my main focus for a while. She was studying to be a lawyer and was very bright, but something was amiss. On one date she insisted on seeing a police movie that both of us knew would be very violent and upsetting. Then, after only ten minutes in the theater, she insisted on leaving. This all seemed rather crazy to me at the time. I'm a fairly spontaneous person, but I was nervous about someone who was even more impulsive.

I tried once again with Miriam and heard the good news that Eric was going to Florida and then back to Norway to farm. So we finally started dating, and I remember distinctly that I didn't know what to do when I dropped her off at her apartment the first time. After an awkward pause, she grabbed me and kissed me. Yay!

We started dating more often during the depth of the Minnesota winter. Once, somehow, the subject of Diane came up and I described how clever and interesting she was. Miriam seemed genuinely hurt by this and talked about her own accomplishments and how smart she was, too. This emotional outburst made me feel bad, but it's when I really started to love her.

Then I got sick. A bad case of mononucleosis hit me and I was out of work for about four weeks. During that whole period Honeywell kept supporting me with calls and paychecks. Dad said, "Never leave that company; anyone else would have fired you by now."

A story I love to tell about this period causes Miriam to cringe. She often stopped by my parents' house to see how I was recovering, and one afternoon while I was on the living room couch I blurted out a marriage proposal. She quickly said yes. I claim that I was sweating and delirious at the time but Miriam disagrees. It happened in March 1973. On April 2 we made it official with a fancy proposal dinner in downtown St. Paul. After a walk in Kellogg Park along the Mississippi, a modest engagement ring we had picked out together was presented, and then we drove back to tell our parents. Somewhere along the way I became panicky. This was really it; telling our parents would make it official and irreversible. I pulled off into a parking lot and just shook for a couple minutes. Finally Miriam said, "What's wrong with you?" So we went and told the parents. My dad's first words were, "What ho!" I guess he was surprised.

All in all, a great decision.

The Times

Understanding the economic mood and environment in 1973 is instructive. The Vietnam War was ending (Paris Peace Accords were signed in January and U.S. troops pulled out by March). Arab OPEC countries began an oil embargo that raised the cost of a barrel of oil from $3 ($18 equivalent today) to about $12 ($72 equivalent). This was in response to the U.S. role in the six-day Yom Kippur War where we supplied arms (including nuclear weapons) to Israel. A stock market crash occurred in 1973 and 45 percent of the market's value was lost. This was all against a backdrop of President Nixon having frozen wages and prices in 1971 and then suddenly reversing himself and letting the international exchange rates float in 1973, causing uncertainties and inflation that would last through the decade.

Very few college-educated young people had been in favor of the war, including me. But the nation was strongly divided on the issue. People my age simply didn't want to work on military projects.

Honeywell attracted new employees by beginning environmental and alternate energy work. Avoiding the use of oil became a national challenge, much like the space program had been in the 1960s. Somehow, I began to work on all three: space programs, military, and oil-avoidance programs. I really wanted to work on solar energy but got thrown into many other areas. My feeling about the military work was that it was generally for defensive purposes, not offensive, but that line was fuzzy. I never worked on bombs, bullets, mines, torpedoes, etc.

Solar Energy Coatings

Contracts were flowing into Honeywell for collecting solar energy, converting it for heating and electrical use, and for storing the energy in eutectic phase-change baths. These eutectic baths retain both the sensible heat and latent heat of the chemical mixture, achieving high storage efficiency. Honeywell's connection to home and building heating systems, thermostats, etc., were a major advantage, and new government funding was easy to find.

Someone suggested that I look into the coatings used in solar collectors in an internal R&D project. IR&D projects were used to develop proprietary technology that would be owned by the company. I wrote a proposal and asked for $5,000 to do the work, which was rejected. Here again, Henry Mar talked to me and said $5,000 was simply too low to be taken seriously. So at the next opportunity I bumped it up to $35,000 and got the money.

I was back in a basement lab again doing electro-chemistry, not unlike my work electro-polishing metals at Illinois or back home in my basement chemistry lab. Working with test tubes and beakers after reading what others had tried, I developed some very special, important coatings called "two-layer black nickel." A little like Edison and his light bulbs, it took hundreds of experiments before the coatings worked, having both high solar absorptivity and very low thermal emissivity. One wants the black coating to absorb the sun but not heat up and radiate it all away. The coatings made solar-thermal collectors significantly more efficient. These two-layer black nickel coatings were unique, and references to papers I published can still be found on Google.

Another factor in solar collector performance is the glass that protects the black surface for thermal collectors. The collectors often used two layers of glass, like multi-pane windows, to prevent heat loss. However, normal panes reflect 8 percent of the sunlight (16 percent total). Since the collector may be only 40 percent efficient, the glass reflectance can cause a 50 percent loss in overall efficiency. I developed some very interesting inexpensive antireflection coatings that didn't require large vacuum deposition chambers. The coatings took advantage of the soda and lime that is in most commercial glass. By putting the glass panes in hexafluorosilicic acid super-saturated with silicon dioxide, i.e. silica, the soda and lime are etched away without affecting the main silica component. The result on the surface of the glass is a layer with a lower refractive index, that is, an antireflection layer. By altering the concentrations and timing, very effective coatings were developed that reduced the reflection by about 90 percent. My coatings greatly increased the potential efficiency of solar-thermal collectors, and Honeywell and government agencies wanted to try them out on a large scale.

We got a string of government contracts to optimize the coatings and to make them durable, which lasted several years, but the real fun was in Honeywell's plating facility. It was a large ventilated room where a half dozen 30-foot lines of equipment put electrochemical coatings on the surfaces of Honeywell products. The room was full of crusty old union guys who had probably worked there for decades. There was an unused area at the end of the room where Honeywell execs said we could scale-up our solar coating process. I'll never forget my first walk across that room when a worker said loudly and sarcastically, "Look who's come to save us."

I started buying large tanks that could hold 3-foot-by-8-foot sheets of steel or panes of glass, where we could duplicate the processes I'd developed in test tubes. Ventilation hoods and power equipment were installed to drive the reactions. Carpenters came in and built scaffolding along the tanks so we could walk above the tank surfaces. Soon, pulley systems were lowering glass or steel into the appropriate tanks and solar collector panels were coming out. Up to six sheets of glass could be lowered in and come out with antireflection coatings at a time.

Eventually, Honeywell received contracts to build thousands of square feet of collectors for a Brooklyn Park school, rooftops in Minneapolis and Washington D.C., and for a mobile laboratory/demonstration vehicle (a mobile home with collectors on its roof). My plating facility would have to move to larger quarters. But first, there were two very interesting events.

Henry Mar always considered me a little crazy when I would walk on the edges of the chemical baths to stir the dangerous liquids with a long broom handle. People started calling me "The Witch Doctor." Then, one day, my chemistry student aide, Ray Lin, was working with the glass-etching bath and put a small amount of water into a container of the dangerous acid, something one should never do because the water can heat up from the dissolving process and start to boil, spattering acid out of the container. Somehow Ray made that mistake and fumes from the hexafluorosilicic acid started oozing into the air. Not only could that acid eat through almost anything but it was also very poisonous. Accordingly, that portion of the solar plating area had been isolated with plastic sheeting so that any fumes not taken out by the main vents wouldn't escape. But the room was filling up with a hazy mist. Ray put on two face masks and two layers of rubber gloves and cut a slit into the plastic sheeting. Reaching through, he poured the overheated container liquid into the vented tanks. Then we all got out of there. The next day, Ray seemed fine, but we surveyed the area and found that all the nearby windows had etched fuzzy surfaces.

A couple of nights later, a mysterious fire started in the wood scaffolding near our plating tanks, destroying their usefulness. The union guys undoubtedly had had enough of the dangerous young scientists invading their territory. But it was time to move out anyway since we could no longer keep up with the demand for solar panels. We transferred the work to a local company, Superior Plating.

The later history of my coatings is instructive. Around 1977 I flew for the first time on a corporate jet with the head of the aerospace business, Jim Renier, to the headquarters of the Lennox Corporation in Marshalltown, Iowa where Dr. Renier sold our ownership of the solar coating patents and technology for several million dollars. Honeywell management had decided it didn't want to be in the large-scale metal-bending business, and the coatings had a better future with a furnace maker. Still later, Lennox sold the coating technology to a small St. Paul business in the University Avenue area, which I visited around 1979. No one in that company knew where the technology had come from, having gone through four handoffs. I only recognized one name that they mentioned. But I was happy to see that about 40 people were making a living off my coatings, working three shifts per day. Unfortunately, only a few years later, it all collapsed when the price of oil dropped. Oil had risen to $35 per barrel during the oil crisis but dropped to under $10 by 1980. The solar energy business effectively disappeared. My learning from all this: don't work on things your company doesn't really care about.

When the solar collector development business was growing rapidly in 1974, the company decided to split it off into its own facility, the Environmental Research Center on the top floors of a multi-story building on Highway 36 in Roseville, Minnesota. They asked if I wanted to join the new group, but I declined. I felt an immense amount of loyalty to Jaan Jurrison, who had hired me, despite all reasonable logic, and I was working with a really fun group of people. So my solar work slowly declined over several years and I started to do research on paints and satellites, which will come up later.

Getting Married

Before our engagement, Miriam and I had seen each other about every other day. After our engagement, we were together whenever we weren't working. We went to every restaurant, movie, play, or symphony available, and even now we miss being as connected to the city as we were during that period. We also played tennis, went swimming and bowling, and rode bicycles 30 miles to my uncle Del's house and back. Courting can be hard work. A lot of the time, though, we just hung out, with Miriam grading papers and me working on Honeywell stuff.

Soon Miriam entered a hyper-organizer mode. We had people to tell, invitations to send, a church and pastors to reserve, a new apartment to rent, furniture to buy, dresses to select, vows to write, showers to attend, and a wedding and honeymoon to plan. Miriam has never left the hyper-organized mode—a great benefit to me.

There were a few scares along the way. Once, when looking for dinner-
ware, I kept referring to ornate, complex designs on plates as being "interesting." Miriam waited a long time before panicking and speaking up.
She thought I liked those plates, and I had to explain that "interesting" just
meant interesting, not good. In fact, we had very similar tastes, both liking
clean modern designs with warm, traditional overtones. We've never had
major disagreements about purchases in our lives.

Renting our new apartment on Rice Street took some effort. If you ever
look into a (now-antique) phone book in Minnesota, you will find page after
page of Petersons. You will find a column and a half of Ronald Petersons
and at least five Ronald E. Petersons. Getting the apartment was difficult
because one of those Ronald E. Petersons was writing bad checks. After a
couple weeks of paperwork and checking, they allowed us to rent at Rosedale
Estates North, which sounds a lot better than the reality. It did have a cool,
sunken living room, a balcony, and a fireplace.

We asked two pastors to preside over the ceremony, one from the large
church where the wedding would take place, Rev. Leman Olsenius, and one
from the small church Miriam attended, Rev. Tom Schultz. I think I waited
until the last minute to write my vows because Miriam wanted a promise to do the dishes in the text. We settled with code words. We secretly
agreed that "I will respect and honor" meant I would do the dishes when she
cooked. Years later we had a testy argument about what exactly was the right
balance of housework between us, and a shift on my part seemed to help. I
agreed to cook on weekends while Miriam cooked during the workweek—
not exactly balanced, but workable.

Many marriages between people of our age have fallen apart, I believe,
because an equitable balance about housework and raising children was
never achieved. Many men assumed they had no responsibility in the home
because they had outside jobs. That particularly made no sense when the
wife also had a career. Rather than talking it through, some would stew inside silently until everything blew up. Miriam still does more family work
than I, but we found a fair ratio, and we love each other.

Our wedding, on September 22, 1973, was a wonderful affair. Steve
Landy was my best man and Lynda Thompson was Miriam's matron of
honor. Steve and I played basketball in the morning to relax while Miriam

did a dozen things at the church, Gustavus Adolphus Lutheran in St. Paul. About 200 people attended the wedding. My sister Janis, Kris Palfe, Bill Cox (Miriam's brother), and Jack Mutchler (cousin Marsha's husband at the time) were all in the wedding party. Miriam's brother George and my cousin Steve were ushers. I was thrilled to walk down the aisle while the church pianist played "Wedding Day at Trollhaugen" by Edvard Grieg. That music was my one contribution to the music schedule. We said our secret vows. We were really married.

At a reception in the church dining hall we signed the documents proving we had been legally married. The reception was a refined, reserved affair with no loud music or alcoholic beverages—very Scandinavian. By 9 p.m. we were off to the Hilton Hotel in downtown St. Paul. My Cougar was decorated with a large Mickey Mouse picture and the hubcaps were filled with rocks, which made a terrible racket as we went up the hotel ramp.

Sometime after everyone had left, we realized that we had forgotten the marriage license at the church. So we drove back to Gustavus Adolphus to look for it. A janitor was still there and helped us dig through the trash barrels for a while. We hoped someone had grabbed the document or that it was rolled up in the paper tablecloths somewhere. We told the janitor to not throw anything away and called our friends and asked them to search the next day. We got back to the hotel very late, fatigued and anxious, and our first full sexual encounter wasn't particularly memorable. We went to sleep worried that after our honeymoon in Mexico, Miriam wouldn't be able to get home because her ticket name didn't match her passport, and the wedding certificate to prove she was legit was missing.

The next morning we flew off to Acapulco and hoped for the best. We only had three days available before Miriam had to get back to her classes, so we booked the most expensive, and highest rated, hotel in Mexico—the Las Brisas. It cost $200 per night at the time, or about $1,200 today. Situated on a high bluff overlooking the lights of Acapulco across a bay, it was a beautiful place. Each guest received the keys to a pink and white striped Jeep and most guest rooms had private outdoor pools.

When we arrived—disaster. They had somehow lost our reservation and I had forgotten to bring along the confirmation number. So we were stuck in a small room on the backside of the facility with no view or pool. Miriam was seriously starting to think that marrying me might have been a big mistake. The super-organized one had married a lame-brain nonplanner. We asked the counter people to please search for our booking in their paperwork.

Fortunately, the next day they found our reservation and we moved to a beautiful individual suite with a gorgeous view. Each morning a man came to remove wilting hibiscus blossoms from our pool and replace them with new ones. Miriam and I sat with exotic drinks in coconut shells enjoying the warm breezes. We slowly got the hang of the sex thing, several times per day.

We got out into town in our pink Jeep, with me shrieking the clutch on the steep hills and narrow roads of the Las Brisas resort. We saw the cliff divers, parasailors, and beaches; swam; and went out for fancy dinners. One night we ate on the top floor at one of the most famous restaurants, El Número Uno. When we left, we found our Jeep surrounded by a bunch of local young men. They demanded in broken English a protection fee for watching our Jeep during dinner. Miriam said something to them in a loud voice and they quickly dispersed. She refused to tell me what she said, but I was very impressed.

A photo taken of us at the Las Brisas Lagoon restaurant shows how amazingly young and innocent we looked at the time. We were in love. Luckily, we got back to the U.S. with no one at the airport passport station questioning Miriam's name. Home in our new apartment, we had the rest of our married lives ahead of us and could prove that we were really married— someone had found the marriage certificate.

Dr. Quack

When the Environmental Research Center finally spun off from S&RC, those remaining moved to a large building on Ridgeway Avenue and reorganization took place. Henry Mar became my section chief in a new coatings section. Henry was my mentor and a true friend. I had many adventures with Henry—most involving food. We often traveled together to possible funding sources for our research. My ability to sell projects stunk. Our marketing man, Leo Fielder, said I was the worst salesman he had ever seen and that he had a lot of work to do to whip me into shape. Technical people have a tendency to try to sell ideas they know or are working on. Gradually I learned to actually listen to the customer and to try to speak coherently.

In 1974, Henry Mar and I attended a weeklong conference in Monterey, California, and toured Pebble Beach, Spyglass, Fisherman's Wharf, and the whole peninsula during breaks in the conference. We actually played golf at Pebble Beach, not a good idea for a duffer like me. I think I stopped counting strokes after 130.

Every time I traveled with Henry he picked out the best restaurant in the city to visit. For one dinner we went to Carmel and searched for a place called the Pink Lady restaurant. We found it, but they said it was booked for the night but we could come back later in case there were cancellations. After walking around town, we returned and had an excellent but very expensive dinner. We topped it off with a dessert called a French Floating Island. There were only four tables in the restaurant, with seating for only two at each. We chatted with a man at the next table who actually wore a monocle with a chain hanging from his suit. As we left to pay, $40 each I recall (about $240 today), we found that our bill had already been paid. The kind man with the monocle must have liked us for some reason.

Once in Chicago we found a famous restaurant in the Arts District. After a great meal, I ordered a raspberry soufflé for desert. It took an extra half hour to prepare the soufflé, but it was worth the wait. It became one of the desserts I later practiced and prepared at home for honored guests. About five tries were needed before I could get the soufflé to rise properly. Its secrets are as poorly understood as quantum foam. Fortunately, I found it tasted good even when it collapsed entirely before eating.

In L.A. we attended an exclusive place called the Green Door. I think that's where I first had a Bananas Foster, another dessert I learned to make on my own, with flames shooting high in our dining room. I call it Bananas Peterson, of course.

One trip to San Francisco had a lasting effect on my reputation. Whenever Henry traveled there, he picked up Chinese vegetables and other ingredients for his wife, Jennifer, that she couldn't easily get in Minneapolis. One time in Chinatown he picked up groceries and then convinced me to bring back a frozen duck on the plane. Back in Minneapolis in the Ridgeway building, I walked down the longest corridor, bumping into several colleagues going the other way. I pulled the frozen lump from my bag and said, "Do you want to see my duck?" They never forgot, the word spread, and for 20 years I was known as Dr. Quack.

The Pentagon Papers

Miriam and I began what seemed like a normal life after the bustle of 1973, with the new job, wedding, and all. We carved Halloween pumpkins, prepared Thanksgiving turkeys and Christmas hams, and enjoyed people, exercise, local entertainment, and working hard. I learned to ski during that winter, often going to Lutsen or Spirit Mountain up north with Honeywell friends, Janis, or cousin Marsha. Reading the book *The U.S. Ski Technique* gave me the basics. I remember falling down the first time and thinking, *The book described this possibility on page 67*. Later that day, doing the snowplow down a steep hill, the skis straightened out and I went plummeting down the hill at full speed. People waiting in a towline dove out of the way and I decided to fall just before going into a streambed. My technique improved eventually.

In April 1974 we decided to take a break and visit Washington D.C. and nearby areas to see the cherry blossoms in bloom. We visited Philadelphia, Mount Vernon, the Smithsonian, watched Congress in session, etc. We even had ten minutes to talk to our congressman, Joe Karth, and have a photo taken in his office. One morning we were signed up to tour the White House, but I got an early call from Honeywell. Roger Heinisch was scheduled to give a talk at the Pentagon but had gotten sick. They wanted to know if I could drop in and give the talk for him. Unfortunately, it was a subject I knew nothing about, but not to worry, his presentation slides had been flown in to the Honeywell office in Rosslyn, Virginia.

So, after our morning tour we went out the White House front door and I caught a cab, leaving Miriam to fend for herself. I arrived at the Honeywell office about noon and was handed a thick manila envelope with two dozen transparent slides inside and some supportive written materials. I furiously scanned through the slides and papers while a Honeywell salesman drove us to the nearby Pentagon. He led me through the turnstiles and checkpoints to enter the building and sat me down in a corridor at 12:45 p.m.

At 1 p.m. sharp we entered a large conference room, set up in auditorium arrangement, and I was introduced to a colonel. "Keep eye contact with him," my Honeywell guide whispered. "He's the one with the money we want."

We sat along a sidewall as the room filled. Suddenly, everyone jumped to their feet at attention, while an entering general said, "At ease." By the time I thought about standing, everyone sat down again. This happened three times, and I actually stood the final time, finally catching on to the drill. After awhile, I was introduced and went to a podium with the Honeywell marketer at the projector. Two generals sat in the front row, their chest medals sparkling and their intense eyes staring directly at me. I couldn't stop looking at them and ignored the colonel with the money in the back row. Somehow I managed to talk coherently about the project proposal without panicking too much.

When we got back to Minnesota, and I went back to work, I found that I was sort of a hero. Apparently reasonable people would have turned down my snap assignment while on vacation. I was crazy, but it worked out okay.

Side Jobs

Our reputation in solar energy had started to spread and it led to many exciting opportunities. I was asked to talk at churches, local groups, and I even flew to Ithaca, New York, to give a talk at a physics afternoon wine and cheese colloquium at Cornell University.

A government Summer Research Study Group asked me to come to La Jolla, California, for a week to examine the military and commercial potential of our specialized solar and laser coatings. The program was housed at a local grade school where I met several famous professors and government officials. Miriam joined me on the trip, and we extended the stay for a vacation, driving to Los Angeles to see Disneyland and Caltech.

On another occasion, a very rich friend of some Honeywell employees asked me to consider joining a spin-off solar company. I was treated royally and we discussed the possibility on a small yacht in the center of Lake Minnetonka. They didn't want anyone to hear their secret plans or see who was plotting. I turned him down, but there were other opportunities where I did consult with others.

For example, a man involved with the founding of the Lear Jet Company thought he had invented a new way to power cars. Honeywell manager Roger Schmidt arranged a time for us to meet with him. The man had a pile of registered patents for his idea, which involved using evaporating liquid air to drive pistons and move an auto. The pictures showed a large radiator sticking up on the roof of the car.

Thermodynamics wasn't my favorite subject, but in my spare time I analyzed the idea (we also were collecting a handsome consulting fee). With very nice diagrams of temperature and pressure, I explained that there wasn't enough potential energy in liquid air to propel a car. The man wouldn't accept no for an answer. "But the liquid air is only 30 cents per gallon and it's completely clean," he complained. Later I came back with a picture. I said, "If the gas tank (liquid air tank) is as big as an elephant, the car can go one block." He fired us, and we later heard he was working with other people. He never saw his car in production.

Our House Is a Very, Very, Very Fine House

With both Miriam and I working full time, we were beginning to accumulate savings. We occasionally graphed the progress of our joint savings account on a large piece of paper. It steadily rose, except for occasional purchases and tax payments. By 1975 we thought we had enough for a down payment on a house and started looking around.

We came close to buying a couple of existing houses and then saw one that we really liked. It was located on a busy street, but the builder started talking to us about building a similar one on an empty lot he owned. Bob Walsh was the builder, and his father was a farmer who owned several acres of land in Shoreview, which Bob was selling off, lot by lot. He showed us available lots in a field by an open slough teeming with wildlife. Ultimately, we snatched up the one we liked best. So, in the summer of 1975, we owned a $7,000 lot.

It would be almost two full years before we moved into the actual home. The lack of a street was the problem. We spent a year going to Shoreview City Council meetings to lobby for the approval of road construction, sew-ers, waterlines, etc. The delay turned out to be helpful because we had time to carefully examine and refine the architectural layout. The whole house was essentially redesigned before construction began. We extended the house by two feet on each end, vaulted the dining room and living room ceilings, added rough-sawn beams, switched a family room for a kitchen, added a formal dining room, and squeezed in a walk-in closet.

I built a small model of the house using cardboard, with tiny peek holes in the walls so we could see what the rooms would look like. Today I would have made a 3D computer model, but that didn't exist then. I even put possible wallpaper choices on the cardboard walls to check them out. Walsh accepted the design changes and quoted the total price of $66,400, which is less than what the lots were selling for 10 years later—a pretty good deal. And then we waited and waited.

Construction began in December 1976 and extended through the winter and spring, with our move-in in April. I wired the whole house for future TV and stereo speaker connections, and my dad and I tiled the front entrance. Bob Walsh was a pretty good builder but a lousy accountant. He was starting to sell many houses and never really finished ours, leaving a few small items undone. As a result, he never submitted his final bill for all the additions and changes we had made, probably saving us several thousands of dollars. Yay!

Our relatives helped us move in and we began to invite friends over to see our new place. We held a telescope party off the kitchen deck and mostly watched the ducks, herons, and other birds close-up with the scope and binoculars. Soon we noticed that deer, pheasants, foxes, raccoons, and other wildlife considered our lot as their home as well. We planted sod, trees, and flowers, and by August the backyard was green. But the construction was not done—an empty, unfinished basement loomed and cried out for action, but many projects demanded equal time.

Photography and Painting

In 1975 we took a wonderful trip, my first to Europe. Landing in Frankfurt, we rode a train to the city center, looked confused, and got directions from a helpful German passerby. My two years of high school and one month of college German were put to the test. We found our way to a nearby rental car lot and took off.

On all our trips over the years, and we've had many, Miriam has acted as travel agent, or at least the person who hooks us up with a guide and buys the tickets. Our first European trip together took us to Heidelberg, Zurich, Luzern, Interlaken, Zermatt, Bern, Oberammergau, Munich, and back to Frankfurt. Switzerland remains one of my favorite countries for its sheer beauty.

Miriam noted how many hotels named Eingang there seemed to be. "Must be a chain," I joked. I was parroting our friends who had visited Germany earlier and didn't know that "Eingang" meant "Entrance." Toward the end of the trip my German did miraculously begin to bubble up, and I was able to order food and make room bookings in German when necessary. It's amazing what's stored for reuse in our brains, waiting for a need to pop out again. On that trip we went to the information booth in each new city and asked them to find us a room, but then I had to do the talking to the owner. We aren't quite as daring anymore.

That first trip was important because it really started my lifelong interest in photography and, later, in painting. I had purchased a new Honeywell Pentax camera for the trip and got some great pictures of mountain scenes, the Munich Olympics area, the Matterhorn, and from an Interlaken Jungfrau train ride. It was also where Miriam and I met the U.S. bobsled team training for the Innsbruck Olympics in the famous Munich Hofbräuhaus, and Miriam was encouraged to drink over a liter of beer—not a good idea.

Back home, I started trying to recreate our trip photos in oil and acrylic paintings. Still later, I attended painting classes at the University of Minnesota to learn how to get better. One very odd professor gave this advice, "Paint what you want, and if there is a part that's bad or boring, fix that part." That was the sum total of his painting lessons. Actually, it's not bad advice for life in general. Toward the end of the classes I got to paint nude women. One was very fat and another very beautiful. I gave my finished painting of the pretty one to my sister Janis' boyfriend, Mike Keenan, at Christmas. He said he would hang it with honor in his garage. Many paintings were given as gifts over the years, and many hang on the walls of our house.

Rainmaker

My work in the Coatings Section of S&RC began to branch out over time, and I developed a reputation for being able to bring in the money—a "rainmaker" in modern parlance. In one contract we tried to develop an inexpensive paint coating to replace the special black nickel coatings for solar collectors. That led to many paint-related contracts, so many that others started referring to it as Peterson's Paint Factory. This was not meant to be complimentary since Honeywell had no real interest in paints, but it brought in a lot of contract money.

One type of paint is still being used on a low-flying troop support aircraft, the A-10. The problem was hand-held missiles, which could pick up the sun glint or infrared signature from normal paints and shoot down the planes. I wrote a proposal to solve the problem using a new paint material that I called Utopium Oxide. The name signified that I didn't know what would work, but would look very hard to find something. Somehow we got a contract from Wright-Patterson Air Force Base to find it.

I hired a new student and decided there was no way to find the material by theory. So we purchased every solid material mentioned in the *Handbook of Chemistry and Physics* that cost less than $10 per gram and that wouldn't explode when exposed to air, about 300 compounds in total. The student was given the summer job to measure the reflectance of every one, a lengthy, mind-numbing procedure. Later we learned she ran off to Egypt and was having schizophrenia issues—hopefully we didn't cause them. In any case, after a whole summer of searching, she found only one oxide with just the right properties: white in the visible, yet black in the near and far infrared. We ground up this powder, added a very rough transparent material (to stop the glint) and put it in a polymer fluid. It worked.

Unfortunately, Honeywell had no interest in making the paint for the Air Force, so the formula was transferred to Deft Chemical. It has gone on hundreds of planes through two or three wars, protecting our pilots.

We also developed a "chameleon paint," which would be green or camouflage colors in cool places but would turn white or light tan in hot areas, like a runway in Iraq. But that paint was never manufactured because it was too expensive.

Along the way I essentially developed a formula for winning contracts
that went like this:

- Visit a contractor company or government agency with a hot new
  idea to get their attention. It will take about two years to get in their
  funding queue—so don't worry about it.
- Then listen to their problems or upcoming contract opportunities
  and find a way to help them.
- Learn the names of their favorite colleges or other contractors.
- Put together a team with all their favorites who could help solve
  their problem.
- Give the approach a catchy name, like "Utopium Oxide" or "Laser
  Eater."
- Write a coherent, exciting proposal that puts your best ideas up
  front (in case they don't read it all).
- And wait.

At one point, we won eight consecutive competitive contracts, which was
rather unprecedented. Each opportunity had about five to seven competitors and the chance of winning was 30 percent at best. The Coating Section
was on a roll.

The next major problem we worked on was protecting satellites from lasers. The U.S. has many important satellites in orbit and the Air Force was
especially worried about someone pointing a powerful laser, or even a not-so-
powerful one, that could damage their optics or other sensitive components. I
learned a lot from our first contract in that area, due to a big mistake I made.

The idea was to develop a coating to put on the satellite optics or window
that would reflect all known powerful laser wavelengths and let the rest of
the light through, so the satellite could still do its job. No one knew if it was
even possible, so I used our unique computer simulations to design one. A
successful design needed over 200 ultrathin layers and we went to work to
actually make samples. After optical testing and near the end of our allotted
two-year contract, we succeeded. I proudly brought the coated samples to
our contractor. Sitting at his desk, he pulled out one of the coated disks and
thanked us for the good work. Before leaving, he blew his breath across its
surface to see if it would fog up, and to my horror the coating we had worked
months to produce curled up off the substrate.

The lesson here was to not just worry about the difficult part of a development. There are dozens of other requirements, like durability, humidity resistance, cost, customer quirks, etc., that all have to work before a product is successful. I never forgot that lesson. Honeywell got a black eye on a $300,000 contract (worth about $1.5 million today) because we didn't have time to do all the tests. I was forced to write a 150-page final report to explain what we'd done and why it failed, which didn't make our customer feel any better. It's much better to succeed in all the contract goals. The reports can be very short and the customer will be happy.

Some U.S. satellites are critical. For example, the Defense Support Program (DSP) satellite looks for intercontinental ballistic missile (ICBM) launches and the Defense Meteorological Satellite Program (DMSP) tracks weather formations. Both have high-performance optics that focus light down onto sensitive optical sensor planes. Even a weak laser hitting those optics could heat up and burn the sensor, so protection was needed. In fact, the government wanted materials and strategies to protect all parts of our satellites, so they started a series of Satellite Materials Hardening Programs (called SMATH), which were very well funded. We won multi-million dollar contracts from McDonald Aircraft, TRW Inc., and directly from the Air Force Space Command to solve these problems. I invented several techniques: (1) an interferometric device called the Laser Eater, (2) a tunable Fabry-Perot optical filter, and (3) a phase-change protective coating, Vanadium Oxide, a compound that is transparent at low temperatures and becomes reflective when heated. The idea was to protect the sensor surface with a coating that reflects light only when a laser hits. The Laser Eater never quite worked, although it did help us win the contract.

The tunable Fabry-Perot filter was my most important invention. In fact, the patent started a squabble among some Honeywell people. I invented it with a colleague, Bill Durand, sketching out the idea on our white board. But there were now so many contracts that I couldn't handle them all, so I turned the tunable filter idea and contract over to another person, Anil Jain, who built it. He later complained that he should get patent credit alone, which Bill and I disputed. Anyway, all three names finally appeared on the official patent. Altogether, I got five patents during this creative part of my career.

It was all about to change.

Janis

While my career was rolling, my sister Janis finished her master's degree in education from Hamline College. I want to include a statement about how proud I am of her. Remember, when she left high school, suffering from Crohn's Disease, she weighed only about 77 pounds. She endured a series of surgeries in which portions of her small intestine were removed. My parents worked extra hard and scrimped to pay for hospital expenses not covered by insurance so that she could go courageously off to college and graduate in 1969. She often quipped that the new wing of St. John's Hospital should be named after her because of all the insurance payments. But through it all she finished her education and degree and got her ideal job as a kindergarten teacher. She was, by all accounts, a great teacher, whom the kids truly loved and learned from. Years later, almost every student remembers her and what she did for them. Some former students are now grandparents.

In some ways we are opposites. I am a true introvert at heart while Janis loves people. She is gentle and keeps in contact with all of our relatives and does kind things without prompting. Everyone knows her and loves her.

We didn't always get along. I'll admit that I occasionally teased her when we shared a bedroom through high school. Frankly, she was and still is a nervous person, easy to annoy. I was a kid. But that doesn't mean we don't love one another.

Janis struggled through a few awkward relationships. Both she and I have difficulty confronting other people. Then, luckily, she met Fred Behrens, who also had never been married. In October 2009, they wed and are living a life of loving mutual support. I am so happy for her and repeat—I am very proud of my sister.

Sweatt Award

In 1977 I won the Sweatt Award, Honeywell's highest technical honor, for my work on solar coatings. The award was named after H. W. Sweatt, one of the founders of the company and it was given each year to about 15 people throughout the company from about 50 nominees.

At the time, the title and role of Technical Fellow hadn't yet been invented for Honeywell, which meant that compensation and respect for technical people had its limits. I had been promoted to Staff Scientist, the highest possible level for a scientist or engineer, and future salary raises would be, at most, a cost-of-living increase plus maybe a percent or two. To make big money you had to become a manager. Later, the Fellow ladder would be available, and one could even become a Corporate Fellow and be paid like a Vice President, but that didn't exist in 1977.

I was very happy to win the Sweatt Award, which included a $3,000 check and two trips. One trip was to the city of Minneapolis for two days (whoo-hoo!) and the other was to a technical conference of one's choosing, anywhere in the world. The Minneapolis part included tours of local factories, technical speeches, and a big dinner with the CEO. The conference trip was the big payoff.

The first question that arose in my mind was, "Does the conference have to be relevant to my work?" The second question was, "Do I actually have to attend the conference?" After some quiet talks with previous winners I learned that the key question was, "Does it look like a business-related trip for tax purposes?" which meant that it should be vaguely relevant to your work and you should be sure to get a receipt to prove you were there.

Accordingly, Miriam and I went to London so I could attend the International Conference on UV and X-Ray Plasmas. There were some optics talks involved, so I figured we were covered. We got a nice room at the Strand Palace Hotel, and on Monday morning I entered the door of the Royal Albert Hall and went to the conference sign-in desk. The reception ladies said I should sign up for the big dinner on Wednesday night, so I paid an extra £5 and said I'd be there. I didn't have the heart to admit that I wouldn't really be attending any of the conference. I picked up my receipt and quietly went out the front door, meeting Miriam across the street in front of Kensington Gardens, and never went back.

We almost didn't make it to London. The night before we were to leave a terrible storm dropped six inches of rain at our house and a retaining wall behind our garage, which I had built, collapsed. There was no way to repair the damage quickly, so we covered the area with tarps and prayed the garage wouldn't sink over the edge before we returned. We had arranged for our friends Carolyn and Bruce Oliver to stay at our house while we were gone. They were probably more worried than we were about the washed-out hill.

Miriam and I soon forgot about home. We were in London, seeing all the sights and having a great vacation. I liked the Strand Hotel so much that I stayed there on several subsequent trips. It was only a few blocks from Trafalgar Square and Covent Gardens. I bought a suit, ties, and a Sherlock Holmes hat at stores along Strand Avenue.

Queen Elizabeth II celebrated her 25th anniversary during our visit and she waved at us, and thousands of others, as her limo drove by. There were colorful banners hanging along streets on all the light posts. We picked a fun week to be in England.

After several days in London, we rented a car and headed toward Scotland, stopping at York and Hadrian's Wall along the way. But Edinburgh was my favorite city. We rode a bus into the central city and loved hearing the accents of little kids on the way. Back then, Edinburgh was a dark, spooky city, with ornate curls, gargoyles, and statues on all the buildings, and most importantly, everything was black. Over 500 years of coal fireplaces had coated the town with black soot that made everything ghostly and mysterious. They have since sandblasted the soot off all the buildings and switched to cleaner heating systems, but back then it was scary and cool, even during the day.

We had told Honeywell that we would be gone for three weeks, so heading back south we caught a car ferry to Amsterdam, drove over northern Germany to Denmark, ferried again to Sweden, and finally to Norway. Norway was almost as beautiful as Switzerland, and we finished the trip with something called "Norway in a Nutshell," which involved every type of transportation through gorgeous mountains and fjords. We spent the final evening watching the sunset over the bay in Bergen before heading home. All in all, it was an unforgettable vacation.

Back in Minnesota, we thanked the Olivers for watching our house and keeping our garage from slipping into the backyard. Our soil in Shoreview is mostly sand if you dig down about six inches. About two or three cubic yards of it had sloughed down a steep hill, an 8-foot drop. Staring at about 8,000 pounds of wet sand, I knew I needed a solution that would last. So I nailed treated, 6-inch wood logs together with railroad spikes, piled them in place, and then poured concrete behind them. Adding pipes from the top would funnel water down to the bottom. I still worried that water pressure might break the wall, so I added a concrete surface layer and covered it with dirt. Then we planted flowers on the wall and hoped for the best. All of this was done without reading any pamphlets or talking to any experts for advice (probably not a good idea), but the wall still stands.

Amy

In December 1977, Miriam became pregnant. Somehow, despite our over-full job workloads, it set off a period of even greater activity. Maybe we thought our house and lives had to be in perfect order before the baby arrived. We were young and could do anything.

I cleared large areas of snow off the slough/pond in our backyard that January for skating and ran a hose from our basement sink all the way out, to make a smooth surface. But the slough was shallow along its edges at that time and as the ice froze, it lifted up and kept making wide cracks. So I kept reflooding it to fill up the cracks. Eventually we had a skating rink with interesting ripples frozen in the surface, not good for skating, but charming.

As it warmed that spring, we planted trees and gardens with help from Miriam's dad, Ken Cox. Her brother, Bill, gave us 10-inch spruce trees to plant, which ultimately grew to about 40 feet in our yard.

I started building brick flower boxes near our front entry in May. Not wanting to take chances about frost damage, I dug out 6-foot-deep footings and reinforced them with rebar every foot. Those flower boxes weren't going anywhere. They had parallelogram shapes and used bricks that matched those on our house front. Learning how to lay bricks was fun and the boxes were soon filled with dirt and planted with marigolds.

Heavy rains in June caused the backyard slough to overflow, slowly rising past the trees and up to our new gardens. Once again we spent evenings at the Shoreview City Council, pleading for them to do something. Fortunately, our neighbors, the Tedeschis, had a house even lower to the pond. Waves were lapping at their basement door. Sandbags helped keep the water back, and the city brought in a large pumping machine to pull water from the pond over a hill to an undeveloped area that drained into Snail Lake. A couple years later, the slough was dredged with large cranes and a permanent drainpipe installed. More streets and new houses were going in, and the dredged black dirt was used to top off those lots. We asked the crane operators to leave an island in the middle of the pond, which has become a sanctuary for nesting egrets, ducks, geese, and heron. They dug right up to our boundary line and straight down 16 feet, resulting in a nice, clean pond officially called Willow Pond for the two or three willows living on its edge.

That summer, Miriam had at least four baby showers to attend. We hung new wallpaper, including some interesting orange giraffe designs in the designated baby room. The room soon filled up with baby paraphernalia and a crib.

On September 1, at 3 a.m., Amy was born. The birth wasn't a smooth one. We had gone to Fairview Riverside Hospital over 48 hours earlier and the staff had used all their tricks to speed things up. I suffered also, being forced to sleep on two hard, straight-back chairs set at an angle so I would fit. Miriam has never been very sympathetic about this.

About four days later we had this tiny, wonderful little person in our home. Amy was baptized on October 29 and already had very strong legs. I had her practicing standing at only one month old, and at two months she "jumped" off the refrigerator—a ritual observed by all our subsequent children and grandchildren.

Amy's most distinguishing characteristic was gigantic eyes, often open and observing intensely. She also was truly cute, if I do say so. At nine months she was walking and at 11 months, running and kicking a ball. We played with her, photographed her, read to her, and doted over her as only first-time parents can do.

My father, Ed, called her "Two Fingers" for the ones she was habitually sucking, a habit that continued into grade school. She quickly became clever, letting us know exactly what she wanted. At two, she said repeatedly as we drove through residential areas, "Sure are a lot of swing sets around here." In the summer of 1981, Grandpa Ken and I built a swing set and a climbing dome for her.

For her first Halloween outing, she was a bunny, Miriam was a flower, and I was a giant bumblebee as we walked the neighborhood. A couple years later, Amy was eating her Halloween candy in alphabetical order and keeping records in a small notebook. That's our Amy.

Section Chief

I never had the goal of becoming a manager. The research scientist job I had was all I'd ever dreamed of doing. But when Roger Heinisch was promoted to Director, my friend and Section Chief, Henry Mar, was promoted to Manager, leaving an open position of Section Chief in what was then called the Electro-Optical Materials section. The position would come with a nice salary bump, but I didn't really feel the need for more money. Despite my father's money trials, Miriam and I were pretty secure.

I applied for the opening because I didn't like the other top candidate and couldn't imagine working for him. That man, from another group, had a reputation for schmoozing and no obvious technical or marketing skills. Also, he was Anil's friend, whom I had fought with about the Tunable Filter patent.

The decision about the Section Chief opening took over six months. A man named Tolly Kizilos was the S&RC's ombudsman and into people management strategies. He has since written on both management and religion. His stated view was, "For once, can we pick someone who is a people person and not just another technical expert?" Tolly was close to the VP of S&RC and demanded that all decisions be made by consensus. So the decision was delayed despite the fact that I had invented or marketed about 80 percent of the section's work. During interviews, I told them that I hadn't considered it my job to know the deep personal circumstances of all the engineers and become everyone's friend, but I probably could. Ultimately, I was given the chance to prove myself. In February 1979, I became a Section Chief. I had a lot to learn.

One of my first decisions was to hire a new section secretary, since Henry's had moved with him to his new job. After interviewing a few, one named Rachel, who was just out of secretarial school at Northwestern College, came to talk to me. I said I'd like to try her out for a week to see how she would do and she responded, "Just hire me or don't hire me. I won't do a trial run." Rachel was a self-assured young lady. So I hired her and she turned out to be perhaps the best S&RC would ever see. The last I heard, she was the CEO's Executive Assistant at Medtronic.

Our financial assistant kept saying I should reward myself for the promotion by buying a new car. My turquoise Cougar was about 13 years old and starting to break down. For years I had been adding new Bondo to the rusty sides and spraying a new turquoise paint coat on the sides. The car was half plastic, and up close it looked furry from the years of spray-can repair. So we bought a yellow Mazda RX-7 to replace the old Cougar, and with Miriam's yellow VW Rabbit, we were a two-yellow-car family. The Rabbit turned out to be a lemon, and Miriam traded it in fairly soon.

Managing people was the real challenge of my new job. Anil left fairly soon to start his own company, not surprising. He became very successful in that new small business, APA Optics.

When another scientist, Tim Board, decided to leave, it was far more painful. Tim and his wife were friends of ours, and his leaving was a surprise. I had taken classes in how to be a supervisor that warned about micromanaging. So I had mostly left Tim alone. I trusted his work completely but, in retrospect, I didn't spend enough time boosting and discussing his efforts. Being a manager requires just the right amount of encouraging or questioning of an employee's work. Eventually I became very good at that, but not so much at the beginning.

Another interesting employee was Ray Lin, the chemist who had helped with the solar-coating lab. Ray found it difficult to ever ask for help, and I found him reading engineering books on electronics when the person in the next office could have solved his problems. Many years later, Ray left Honeywell to start a chain of Chinese restaurants that were excellent. The last I heard, he was working for an optical company again.

Another person in our group was a perpetual pessimist and complainer. His attitude dragged everyone around him into a state of funk. We endured his gripes until he left a couple years later. I imagine now that I should have talked to him and set boundaries, but at the time ignoring him seemed best.

One final person to mention is Jim Lenz, a new hire in my group with an M.S. in physics from the University of Wisconsin. We had a head-to-head. Jim turned in a final report that was gibberish, in my opinion. He said it was my problem if I couldn't understand the report. I replied that I was smart enough to understand anything coherent, that the report was his problem, and that it needed to be rewritten. I reminded him that I was his boss. He did it again, and we eventually became close friends. We worked together on many subsequent projects. He became a true sensors expert, and eventually worked as Section Head, taught at the University of Minnesota, and was recently Manager of Enterprise Sensors for John Deere.

During my years as Section Chief, I was asked to attend special training. A week at St. John's University taught me that I was quiet and introverted but that I had dreams of exhilaration and power. When asked to write stories, I wrote about spies, presidents, and heroic actions. Taking a Myers-Briggs test there showed that I had an INTJ personality—introverted, intuitive, thinking, and judgmental—the ideal personality for a scientist or one who manages scientists—not a bad prediction. I also learned how successful teams needed to be balanced, structured, and renewed occasionally with new people.

At a training class in Maine called Basic Human Interaction, I learned even more about myself and how to get closer to others. In my time as Section Chief, I learned that each person is different and requires kindness and careful listening in order to motivate or inspire them. Who knew?

All the encounters, problems, or successes with people in my group during this time were crucial to any future success I had. Being an introvert, I spun this experience about people and teams into my daydreams and spare-time thoughts, trying to understand, absorb, and employ what I had learned.

Galilee

As I mentioned, I completely ignored church and religion during my time at Caltech and Illinois. Before college, I hadn't really appreciated the services or sermons at my mother's church, First Lutheran in St. Paul, which seemed rather boring. I did, however, have a pretty good Bible education. My dad was a Seventh Day Adventist and his brother Henry was a national official in that church. Henry had sold us an illustrated, 14-book, encyclopedia-like children's version of the Bible that I read cover to cover. I was confirmed and baptized at about age 15 after two years of classes about Jesus and Luther. But by 1973, after years of religious abstinence in college, I had grown very skeptical and uninterested in the whole thing.

When Miriam invited me to go to Galilee Lutheran Church in Roseville after we were married, I was reluctant. I told her I would go but that I reserved the right to sleep-in on Sundays when I wanted to. Miriam was one of the very first female council members at the small church and invited me to all the fellowship events as well. The only thing I liked less than church services were social events with a bunch of people I didn't know. They did have fun games in people's basements, and eventually I considered it a challenge to memorize as many names of the people as I could.

Slowly I became more involved with the Galilee church and people, until in the late '70s I actually agreed to be a Sunday school teacher. Teaching forces you to know and understand what you are talking about, and there was a responsibility to teach the kids about God and Jesus. So I became even more connected to the church.

Surprise! It was fun. Mostly I had teens in the Sunday classes and I enjoyed discussions and challenging them. We made movies about Bible stories, about the Passion Week, about Moses and the Exodus, and about the Christmas story. On warm days we walked to the nearby Dairy Queen while learning the Ten Commandments. To interest the boys more, we built wooden items we called the Mangy Manger and the Unstable Stable that were used in Christmas programs.

In 1979, at a Sunday service, everyone was shocked to learn that Galilee's pastor, Lee Pierson, had died in an early morning motorcycle accident. He had been counseling drug addicts through the night and hit a median on his ride home at 5 a.m. People hugged and prayed and talked about him instead of the usual service, which made me feel closer to the small gathering of people and more committed to its mission.

When Amy was about two, I actually gave a 10-minute sermon for the whole congregation. Using large charts as visual aids, I explained three possible philosophies about relationships: the two-year-old view ("It's mine!"); the normal barter approach between adults ("You scratch my back and I'll scratch yours"); and Jesus's way ("unconditional love").

There were actually people in that church, like Lois LaVon Brase, who helped one another with absolutely no expectation of payback or reward. I had never experienced such altruistic behavior before. Slowly I was becoming a functioning, useful member of Galilee and was later selected for the Church Council. But my personal faith was always scattered with large doses of doubt. I never thought too hard about whether it was all true. That would take another couple of decades.

Torture in Mexico

Miriam often took her students on excursions to Spain or Mexico. In 1981 I agreed to go along to help chaperone the students and take photos. Amy stayed in Minnesota for five days with each set of grandparents during the 10-day trip. The Mexico trip was interesting, but for me, horrible and painful.

We landed first in Mexico City. On a hotel balcony I had to explain to one Minnesota girl what the "thing" was on the horizon. "That is a mountain," I said. She replied, "Wow!" But I wasn't feeling well. The 7,200-foot altitude of Mexico City had given me a splitting headache that lasted several days.

Then I fell victim to Montezuma's Revenge just as we got on a bus to travel across the country on curvy roads. We stopped at the silver city of Taxco in time for me to throw up.

The final days were in Acapulco, where I was sunburned over my entire back. It was so painful that I couldn't sleep. We flew back to Minnesota with me vowing never to return. I did return, but not for many years.

Building

Amy's birth triggered an intense building period around our house that would finish off the basement by 1982. I designed and anchored 2-foot-by-4-foot walls to the open basement floor and, after studying Minnesota electrical requirements, wired the whole area for 15 amp circuits. We passed the city inspection.

We told Amy to never put her hand near the open sockets because they were dangerous. So the inquisitive one decided to put a nail into a socket to see what would happen. I saw her buzzing from the current, but she had already pulled her hand away. After that, Miriam insisted that every electrical outlet in our house have plastic inserts, even in areas a child couldn't possibly reach.

We created many rooms in the basement—a shop, laundry area, a large playroom for Amy, a family room, bathroom, a working den/library, storage room, and a cedar closet under the stairs. The details took a lot of time, such as the climbing loft in the play area with a desk and ship-like portals through the walls. We put a cork wall in that room to hang all of Amy's drawings. I painted the laundry room with a huge arrow pointing from the washer to the dryer, in case Miriam got them confused. The new shop got racing stripes. A deep-blue carpet was laid across three rooms, and ceiling lighting and tiles were hung. Built-in bookshelves were added to the den. Grandpas Ken and Ed helped with sheet rocking and Amy helped paint. I worked on everything, including the plumbing and tiling of the bathroom, except installing the toilet. I didn't want any trouble there.

After about three straight years of nighttime building I finally cracked. I just couldn't do any more. Some trim that had been stained was put in the storage room, unused. Several drawers that had sliders attached were never finished. I'd proven I could build just about anything, but simply couldn't do any more work. After that, we hired people to do some of the necessary tasks around the house. Perhaps our greatest hire was Joyce Thompson, a house-cleaner, who has been our support and friend for 39 years now. She started just after Miriam went back to work after Amy's birth. Joyce was the best Christmas gift I ever gave to Miriam (or myself).

One thing I continued to build for several winters was a marvelous ice skating area, which got more elaborate each year. Since the pond had been deepened, the ice was smooth. A snow blower was used to make a wide speed-skating oval around the outer edge. I piled the snow in paths, creating an ice maze between figure skating areas. Using snow, I built chairs and sofas on the ice, and sprayed them from a hose to freeze the outer surface.

Since there were still open lots around the pond, many people with snow-mobiles would ride over the pond, engines roaring, do spins and generally pack down the snow. This really bugged me, because it wrecked the skating areas. I had nightmares that they would turn the pond into a snow-mobile racing arena and build stands and stadium lighting in our backyard. Fortunately, Bob Walsh and others were building homes around us rapidly, and within a few years there were no easy entrances to the pond and my nightmares faded.

Steve's Wedding

Amy started traveling with us quite early. She flew to Oregon for one trip, and we rode together to Indiana to be in my best friend's wedding. Steve

Landy married Mary outdoors at a family farmyard on a beautiful day in June of 1980. Steve had become a physics teacher at a suburban Indianapolis high school and Mary was to become a psychiatrist. I served as his best man.

Amy saw cows, horses, and tractors for the first time. It was a wonderful day. I especially remember Mary and Steve opening wedding gifts out on the lawn. Steve's mom and siblings were there, and his mother's gift was in a large box. Steve opened it reluctantly and found one of the early Apple II computers inside. "Why didn't you just buy me a car?" he said to his mother. An Apple II at that time cost around $1,200, about $6,000 today. I was very impressed.

Good Ideas

Back at S&RC, the Vice President, William (Bill) T. Sackett, periodically invited all the Section Chiefs for open forum discussions. At one, he asked us to send him a quantitative measure of the effectiveness of all our research, posing it as a challenge for the next meeting. As usual, I thought of many possibilities—dollars made by the business units, contracts won, long-term payoff, etc. I had it all laid out on a giant table, and then decided it was all too complicated. So I sent Sackett a simple plan, that we should track "good ideas." After all, that's what a research center is all about. A good idea was something that was truly novel and that had an obvious payoff. Admittedly, it was subjective, but simple enough to be actually used.

So I started applying that measure in my section and walked around saying, "Any new good ideas today?" It sort of became my trademark. Everyone knew I would ask, so they started looking for responses and actually inventing things. What you measure is what you will get. Dr. Sackett liked the idea, because it was simple. He had received a couple dozen measures from the other section heads, but liked mine.

I continued to use my "good idea" measure in later stages of my career with high payoff.

Managing

By July 1982 I was promoted again, to Manager of the Electro-Optics and Sensors Science Area. Thirty engineers or scientists, ten students, and several support personnel now reported to me in three sections: Sensors, Electro-Optics, and Laser Gyros. This was a step jump, a promotion and movement to a new area, and a very significant vote of confidence. The whole group celebrated and welcomed me with a giant cake with a Donald Duck image in the icing on the top. "Congratulations, Dr. Quack," it said.

Managing the Laser Gyro section was very important because it had become a major product line for Honeywell, generating about $300,000 for each Boeing airplane that was built. Joe Killpatrick had developed the laser gyro and had been manager of the area I was moving into. Joe had a reputation, however, of not being a great financial manager, so he received the new title of Research Fellow, a position of honor without the responsibility of managing others.

The financial administrator for the new Science Area was surprised that I actually seemed to care about the business health of the group. There were three types of work available: work on contracts (government or from other companies), internal research or development (paid for by Honeywell corporate), or burden work (work on necessary but unfunded tasks or simply to pay for people without anything to do). It was best to not use burden funds because they caused the overall rates for our work to go up, making the organization less competitive. To manage the finances, one needs to shift people around to keep them on paid jobs, and plan ahead for the times contracts would end. I developed all sorts of VisiCalc documents and graphs to do this well. VisiCalc was the first killer app for Apple Computer (similar to Microsoft Excel today), which turned the Apple II from a toy for hobbyists to a serious business tool. Pretty soon the financial health of the Science Area turned around.

I hadn't forgotten that managing people was also important, so early on I created a questionnaire, about 15 pages long, to survey the feelings, needs, and ideas of all the people in the Science Area. A high percentage actually turned them in (this was before everyone did surveys and people started to hate filling them out). I summarized all the responses and shared the results with everyone in a large group meeting. Then we actually used their ideas to manage the group—and it worked. Also, I systematically spent face time with all the people to know them better.

In small groups we discussed how the internal funds were being spent.

For example, one fiber optic sensor scientist was using his $100,000 of corporate money to try to develop six to eight types of sensors. Just talking logically about this convinced him to narrow that down to three developments, one being a fiber optic gyro—something that paid off well a few years later.

In the Science Area, I also broadened my tabulation of good ideas, constantly asking individuals what they had dreamed up lately. Something was learned from this study that changed my whole view of management, development, and later, even my politics. In my tables, I found that the generation of one of my "Standard Good Ideas" took about $50,000 when people were working on government contracts, about $15,000 while on Honeywell Internal Research and Development (IR&D), and only around $5,000 while working on competitive proposals for new work. The element of competition made people far more creative, an idea I exploited successfully throughout my career.

Two efforts to hire individuals were interesting. One was Howard French, a Caltech grad and astrophysicist who grew to be a close friend. His first visit to Honeywell looking for a job was a disaster. He was caught in traffic on his way to the interview and was late. He spent much of the day apologizing for being late and then describing why he wouldn't be a good person to hire. Needless to say, he didn't get a job offer, which caused him to find work at Sperry Univac, where he also found his future wife, Jo Ann. So it worked out well. I saw Howard again when he joined Galilee church and convinced him to reapply at Honeywell, without showing so much excessive humility. I knew he would be an excellent long-term scientist, and it turned out to be so, with Howard retiring from Honeywell in 2016.

Another important hire was Glen Sanders. Glen was an MIT doctoral candidate when we first interviewed him, which made it easier to keep him away from archrival Litton Aerospace later. Glen ultimately led the development of the fiber gyro for Honeywell.

I need to explain a little about how laser gyros and fiber optic gyros work. In both cases, light is introduced two ways into a closed circuit and travels in loops in opposite directions before popping out again. If the device is stationary, not much happens. But if it is rotating, one path becomes longer and the exiting light beams interfere with each other and come out, either strengthened or weakened, in optical blips. The frequency of the blips tells you how fast the gyro is rotating and hence how much you have turned. Using the laser gyro, a plane can fly across the country and know its direction so accurately that it ends up within a hundred feet of its planned landing strip. This was much better than the mechanical, spinning gyros that had been used previously.

My early time as a manager was going well, but major life events were on the horizon.

My Father's Death

Eddie, my dad, had his first heart attack in 1972 at the age of 56. While visiting relatives he had chest pains and was taken to a hospital in Thief River Falls, Minnesota. His hospitalization served as an informal family reunion, with most of his northern siblings rushing to be with him. Janis and my mom flew to Thief River, and I drove straight through from grad school in Urbana, Illinois, about 14 hours. Modern techniques, like stents, hadn't been invented yet, so he spent several weeks at the hospital and then several months recovering at home. He also had prostate cancer in his early 60s that required surgery. After retiring in 1980 at the age of 63, he finally had time to enjoy life for a while.

Shoveling snow at his home in 1982 caused a second attack and he was rushed to St. John's hospital. He told my mom that he was really scared before tubes and other procedures prevented him from speaking further. I came from work about 4 p.m. and the Holmgrens (Aunt Ruby, Uncle Carl, and cousins Marsha and Debby) were there with Janis, Miriam, and my mom. We spent several hours at his side while the doctors tried multiple techniques to get his heart to beat on its own, but too much of the heart had been damaged. The doctors finally said they planned to remove the oxygen and other life support. I told Dad that Miriam was pregnant and we expected another grandchild for him. I hoped the news might help him rally a bit, or at least that he should know, and his eyes widened so I know he understood. But at 7:20 p.m. we were all there as his gasping breaths got less frequent and eventually stopped. I cried for a long time.

Perhaps every death of someone close to you is difficult, but I think the early death of a parent is the hardest. For years I felt a huge hole, a vacuum for the part of my life my father had filled. Certain songs that reminded me of him caused tears to well up for several years.

My mother lived much longer, and her death seemed more natural. But my father died at only 65 years, not long after his retirement. He was just beginning to enjoy his woodshop and all his hobbies. I blame the stress of his work at Central Warehouse and the bad genes from his mother's side for the early loss. It seemed unfair.

Kelsey Arrives

We were in Cannon Falls, Minnesota, for a reunion with my mother's relatives as Miriam approached her due date. Amy impressed everyone with her ability to read at age four, but Miriam looked worried about being so far from the hospital when early labor signs began. Luckily, Kelsey held back, but not for long.

Perhaps there is a balance to life: one person leaves and another steps in to fill the gap. I felt that way about my father's death and the arrival of our second daughter, Kelsey. Unlike Amy, Kelsey arrived in a rush the morning after the reunion.

Miriam woke me about 5 a.m. and said we should head to the hospital. We drove first to Miriam's parents' townhouse to drop off Amy. While Helen took Amy in, Ken kept trying to describe a shortcut through side streets he used to save on gas. When Miriam got back in the car, I totally ignored his advice and headed to the 35E interstate freeway. Going about 80 mph, a police car pulled us over about two miles later. Today, you're not supposed to get out of your car when pulled over, but back in 1982 I hopped out and waved my arms while walking toward the police car, shouting, "We're having a baby! We're having a baby!"

The cop looked in the car at Miriam stretched out on her seat and asked, "Can she make it? Which hospital?"

I said, "I don't know, maybe. Bethesda."

"Follow me!" he yelled.

Soon we were going down 35E at 90 mph with semi-trailers pulling off

to the side of the road like the parting of the Red Sea. It was very cool. I told
Miriam to try the breathing exercises. She replied, "We're way past that."

A wheelchair was waiting at the emergency entrance of Bethesda Hospital
(another nice benefit of the police escort). Miriam screamed and cringed
when the wheelchair bumped into an elevator, and she flopped onto her bed
only 15 minutes before Kelsey's birth. Miriam's official doctor never arrived in time, but a substitute helped her through the final minutes. Kelsey,
weighing just under seven pounds, rushed into the world and has been
sprinting ever since.

Kelsey jumped (or fell) off the top of the refrigerator at only three months
while pictures were taken, which was becoming a tradition in our family.
I claimed it helped develop trust of her father, but mostly the babies just
look cool sitting up there. Kelsey was physically strong and was baptized at
Galilee on November 28.

Kelsey developed colic, which made her cry a lot and seem grumpy and
fussy. When she was only six months old the whole family took a weeklong vacation in Florida, first to a hotel on a beach in Clearwater, then to
Busch Gardens and finally to Disney World. Amy loved Disney World. She
also had amazing spatial sense and memory, as well as self-sufficiency. In
the Disney parking lot, every day she would run off to the parked car several blocks away and beyond our sight. She was always waiting there by the
car when we caught up. But the important thing was that Kelsey got over
her colic at Disney World. It just went away, and the happy, smiling, clever
Kelsey we know today burst out.

A couple more stories about Kelsey's early years are very memorable to
me. She started walking in July 1983 at 10 months old. About then, I started
taking her for rides in a carrier on the back of my bicycle. Amy had recently
gotten a new Strawberry Shortcake bicycle of her own and learned how to
ride quickly. Remembering a crash I had had on a steep hill once, I thought
Amy needed to learn how to deal with hills. So Kelsey was strapped into my
carrier and we all rode to the Dale Street hill, about two blocks from home.
At the top, I reminded Amy about using her brakes and she started down the
hill with Kelsey and me following behind. Unfortunately, Amy forgot about
the brakes and started wobbling back and forth wildly, going faster and
faster, and eventually crashed. I put my kickstand down and rushed across

the street to see if she was okay. That's when my bike tipped and Kelsey, strapped in, slammed to the asphalt. Just then a car turned up the hill coming toward us. The driver stopped and came over to help. Amy was bleeding and Kelsey was crying. The lady in the car drove us all home, leaving the bikes to be picked up later. Back at the house, Miriam yelled at me . . . a lot.

Despite refrigerator jumps and bike crashes, Kelsey grew more wily, funny, and spunky every year. Amy used to play teacher for Kelsey and the Schornstein girls from next door, forcing them to listen and follow her instructions. Kelsey thought she was bossy. When photos were taken, Kelsey couldn't resist hamming it up.

Throughout Amy and Kelsey's early years, I used lessons from my Caltech Psychology of Child Development course. The basic lesson: love your children unconditionally through at least the first two years, then slowly start discipline and teaching about life after they know you are totally on their side. The plan worked great. In my opinion, Amy and Kelsey were wonderful children and became splendid adults.

Both children kept demanding that we get them pets, preferably dogs or cats. We resisted. In general, Miriam and I have avoided any commitments that would limit our freedom. That's why we have never bought cabins, boats, second homes, motor homes, etc., even though they were easily affordable. Miriam especially didn't want to take care of pets while I was out of town.

Goldfish and hamsters were the only pets that made the cut. Through the '80s we had a series of hamsters, the most famous being Fuzzball (a.k.a. Fuzzy). We had been promised at the store that all the hamsters were male. But one afternoon Kelsey and I discovered eight babies, looking like little jellybeans, hiding under Fuzzball. We ran to Miriam and Amy in the kitchen with me yelling, "It's a miracle! It's a miracle!" The women in our house didn't accept my miracle birth idea. Later, Kelsey sadly had to watch the babies slowly disappear. One by one, the momma decided to eat her children until none were left.

On July 17, 1989, Fuzzball died. Neighborhood kids came over for the official eulogies and burial in our backyard. Fuzzy was buried in a shoebox and had a small tombstone with RIP FUZZBALL written with indelible ink. Some cried.

Miriam and Her Parents

As I am writing this, Miriam is out scraping frozen snow and ice from our driveway. I'm still in PJs, but she's out shoveling. Within a couple of weeks of Kelsey's birth, she was out raking all the leaves in our backyard with Kelsey in a pouch on her chest. This was not unusual—she's a workhorse, a rock, and a pillar. I am exceedingly fortunate to have her at my side.

It's said that there are five ways of expressing love:

- Words of affirmation
- Acts of service
- Receiving or giving gifts
- Quality time and conversation
- Physical touch

For Miriam, actions speak louder than words. Acts of service make her happy, and they are how she shows affection to me and everyone else she cares about. For me, physical contact and giving gifts are also important, but for both of us, working together on a project makes us happy and close. Miriam is very similar to her mom and dad, so it's important to know a little about them.

Helen Pierson grew up on the East Side of St. Paul. She remembered horses pulling carts along her street, a gas lamplighter walking by each night, and the big sign in her house's window telling the iceman how big a chunk he should bring in for their icebox. Her mother was very strict, but Helen found ways around her demands. She twirled around while ice skating and imagined that she was dancing, one of her mother's forbidden activities.

She met Ken Cox at Glacier National Park on a train trip to visit a friend in Seattle. Ken had grown up on a tobacco farm in Ohio and was serving in the Navy. He had traveled the world, but the Navy was hard for him personally because, unlike many of the other sailors, he actually had firm morals and didn't smoke or drink. He had just received a Dear John letter from a girlfriend back in Ohio and was happy to meet a nice gal from Minnesota. They wrote for years, and in one letter, he proposed. Helen had, by chance, asked her mother to take a look at just one of Ken's letters to see what a nice guy he was, and that proposal letter just happened to be the one they opened and read. When she saw that, Helen said, "Wow!" I don't know what her mother thought.

Before their wedding, Ken stayed in Minnesota for a few days and greatly impressed Helen's dad, a railroad worker, by shoveling not only his long sidewalk but also the entire street in front of their house. He had passed the test. After four years and only a few days together, Helen and Ken were married.

By modern standards, Ken was peculiarly polite. Given the chance, he would use the words, "Please," "Thank you," "May I help," and "Excuse me," all in the same sentence. In recent years, he and Miriam fought about who should carry the luggage to the car on trips. The idea about helping others was so deep in their beings that they asked for and expected nothing for themselves. This made it very difficult to buy Christmas gifts.

Miriam took three months off when Amy was born and then taught Spanish half-time at Harding High School. When Kelsey arrived, she took two years off to be with the children before returning, but then had to move to Como Park High School. By all accounts she was a great teacher. We rarely went to any event without Miriam bringing along papers to correct and grade. In September, the students thought her a stern disciplinarian, but by December, while making piñatas, they loved her.

Miriam brings balance to our family and lives. I was driven to compete, achieve success, make money, and could be a little selfish. That isn't the core of what I am, but it was valuable in my working years. My deeper goal is to invent things, discover, and make the world a better place. Miriam's goal was to teach and serve children and help the future that way. Other than the one brief spat mentioned earlier, and a rebalancing of housework, our lives have been in balance.

On Becoming a Research Director

Some people in corporate management probably spend a lot of their time thinking and working toward their next promotion. Generally, I just focused on doing a good job within the Science Area that was my responsibility. We were doing well financially and some great new product developments were underway. One, a small one-inch size laser gyro, would prove to be a major product line for the company about ten years later. But, increasingly, I needed to learn how to cooperate, compete, and work with people from other groups.

When promoted to Manager in 1982, I was thrown into a world of very aggressive people. The Research Management Team consisted of six Science Area managers, the Research Fellow, and the Director of Research. Sometimes it became a battle of who could talk the loudest, interrupt more, and be heard over other people. The "team" consisted of the managers of Controls, Microelectronics, Sensors (my old group), Computers, and Signal Processing, as well as Joe Kilpatrick (the Research Fellow) and Mel Geokezas (Research Director). I rarely got a word in during team meetings, generally holding back until they wore each other out. Then I would add a humble correction or impassioned comment at the end. It wasn't the way to earn their respect. Gradually, I learned to throw in comments right from the beginning, even if they were minor, contributing and getting involved in the process, "owning the solution," not just offering my vote at the end. As an introvert at heart, I work on remembering this lesson even today.

For teams to be effective there must be a balance among four personality types (I learned this in the Advanced Program for Managers, one of Honeywell's management training efforts). There are authority people who like to drive the decisions, unifiers who care more for the feelings of others, problem solvers who just need a target to work on, and visionaries who set the general directions. In my view, the research team was made up entirely of problem solvers, which might not be surprising for engineering people. With time, I evolved into the missing link, the visionary. Maybe as a scientist it was natural for me to think several steps ahead, to proactively imagine the future and try new things, but it gave the research team a new balance and effectiveness.

A day came when the S&RC would need a new Research Director. Roger Heinisch, the Vice President, had moved to a product division, and John Dehne had been promoted to replace Heinisch. Unspoken was a feeling that Mel Geokezas was not a good candidate to be a future Vice President, so a new, second Research Director was needed and Mel was to chair the selection committee. Every current research manager applied for the job—it was going to be a free-for-all.

I decided to give it my best shot. Even with a strong technical and financial track record, most people considered me too passive for such a critical position. At one research team meeting, Mel asked each of us why we should get the job, which included questioning and critiquing each other—a very strange meeting among all the candidates. My quietness came up. Eventually, I told the story about my parents and how they had argued and yelled about minor things, like how the food was cooked. I said that I hated their fights and decided to never act that way. It was difficult to talk about something so personal in front of everyone, but they understood me better. A little later Mel Geokezas offered me the job.

Then I got an unexpected phone call. Rick Bernal, the Vice President at Honeywell's Corporate Research Center (CRC) in Bloomington called and asked to meet me at a Perkins for breakfast the next morning. He offered me the job as Research Director at Honeywell's other major lab. S&RC focused mostly on the aerospace and defense businesses while CRC's focus was on the company as a whole. CRC had wonderful facilities and a more long-range charter, so the offer was very intriguing. But soon the word about the offer got back to Mel Geokezas at S&RC. Mel substantially bumped up the salary he was offering. Generally, I wouldn't recommend getting oneself into a bidding war this way, never more than once a career, but it worked out well. I took the job at S&RC in March 1985.

The S&RC research labs had grown a lot since 1980. Mel and I split the six Science Areas with about 250 people each. Sometimes I think my rapid promotions were mostly a matter of luck. I was born at the front of the wave of baby boomers. The Honeywell job growth during the 1980s was phenomenal (maybe the Reagan defense push helped). At one time, over 60 percent of the supervisors at S&RC had been at their jobs less than six months. They needed management and leaders. Being a PhD physicist helped, and my Caltech education meant that I could understand almost anything—software, human factors, microelectronics, etc. I was a generalist. The fact that my work at Honeywell, right from the start, had nothing to do with my research at Illinois was a blessing because I was forced to learn and succeed in many new areas. One can become an expert in almost anything in about two years. My new job as Research Director would force me to learn how to be a leader.

On the Job as Research Director

One of my first new tasks was to serve as the butt of jokes at the Center's annual picnic and softball game. Henry Mar, Gunter Stein (our Corporate Fellow), and I had to perch over the dunk tank and endure sarcasm until someone hit the target and we dropped into the water. I also still played in the softball game at that time. The culture at S&RC was pretty unique within Honeywell and across a lot of companies. All the engineers and scientists had individual offices that could be decorated as one liked. One long-term scientist had no furniture in his space; he just worked off a mat on the floor. At an Engineers' Night party, which celebrated the accomplishments of the Center each year, that scientist had won a Sweatt Award and came to the front to accept his award from Ward Wheaton, the stern Executive Vice President of the entire Aerospace and Defense Business Sector. The scientist (whose name I can't remember now) wore a stylish tuxedo with tennis shoes. He looked Ward Wheaton in the eye upon reaching the stage and told him, "I own the tux, but the tennis shoes are rented."

The culture at S&RC went beyond a few incidents or individuals; we were friends there, proud of one another, and all the support people were part of the team. We generally won three Sweatt Awards each year, out of about 15 across Honeywell, or 20 percent of the total technical creation of 50,000 employees.

I've mentioned Joe Kilpatrick, developer of the ring laser gyro, but you should know Gunter Stein as well, our Corporate Fellow. He worked full time at S&RC but also was an Adjunct Professor of Electrical Engineering at MIT. He is credited with developing the theory of robust multivariable control, a major breakthrough in the field. He attracted a continuing stream of MIT engineers to Honeywell to work on the most difficult control problems.

One challenge was controlling the reentry of the space shuttle. Our control group is credited with preventing the early space shuttles from burning up, as demonstrated by a giant thank-you painting from NASA hanging on S&RC's walls. Our control team also worked behind closed doors on flight controls for the naturally unstable, fly-by-wire Stealth fighter and bomber. Gunter also did an important study on the causes of the Chernobyl Nuclear Plant disaster, which were partly due to errors in their control system.

I believe S&RC was one of the top ten corporate research laboratories in the country during its heyday in the 1980s and 1990s. One important internal strategy was to constantly encourage innovation, particularly through our Initiatives Program. Gunter Stein headed up a committee that evaluated new ideas from individuals and awarded several small grants, $15,000 each, for people to try out their new ideas. The total amount gradually increased to over $500,000 per year and even included special Super Initiative awards of about $100,000. Being on that committee was one of my greatest joys as Director. Later, I would try similar strategies on a company-wide level.

As Research Director I needed to transition from Manager to Leader, as I had learned in various training courses. Here is a picture of just some of the key differences:

MANAGER	LEADER
Seeks stability	Seeks change
Plans around constraints	Sets direction
Stort-term	Visionary
The head	The heart
Reactive	Proactive
Directs subordinates	Influences many followers
Does the thing right	Does the right thing

As an introvert, I wasn't going to become a leader through sheer charisma. I had to influence people by expressing logic and showing that I understood, respected, and cared about them. I talked openly both about Center and Company successes and problems, and continued to walk the hallways and drop in to talk to individuals. Understanding and influencing others increasingly meant people outside the research lab—those in business units, corporate leadership, and even other companies. Empathy was probably my greatest strength in the new job. Perhaps my being an introvert actually helped with that.

There were many things to learn. Honeywell had excellent training programs for its managers. Professors from the University of Missouri taught a four-week course, Advanced Program for Managers, that I completed. Then, selected Director-level people participated in a Harvard-taught mini-MBA program, a shortened version of their summer MBA program. They condensed the program to five weekly sessions for Honeywell and taught it at the Philip Exeter Academy in New Hampshire and, later, at a retreat center in Owatonna, Minnesota. I learned a lot.

About that time, S&RC moved into its new building. I had been deeply involved in the design and execution of that new facility, a 200,000-square-foot beauty on Technology Drive in Minneapolis just south of Fridley on the east banks of the Mississippi. Known as the Camden facility, each science area had designed their labs. An example was the ten-foot thick, 40-foot-diameter, donut-shaped slab of steel-reinforced concrete poured for the inertial sensors people. The slab was built around a building pillar and floated on sand and other soft substances to keep it from moving when trains rolled by. It was the largest concrete pour in Minnesota that year. I enjoyed the design collaboration with so many people. Having designed architectural plans for houses when I was about 10 years old, I found designing a whole real building fun.

Maintaining the S&RC tradition of personal offices for engineers and scientists was a major point of contention in the design. It, obviously, increased the overall cost, but John Dehne, our Vice President, fought hard for that standard. Quiet personal space was valuable for inventors and an important recruiting benefit. Great labs, including many "black areas" for top-secret programs, were scattered on the bottom two floors. Soon all the technical people adjusted and thrived in the new building, developing new traditions, including paper-airplane drops from the edge of the three-story atrium, competing for longest and most accurate flight. One time I won, which bugged the flight control people a lot.

In November 1986 Honeywell made a bold purchase, for $1 billion, of the Sperry Aerospace business. I had served on Honeywell's acquisition team that evaluated the purchase, with my time focused on their patent portfolio. Sperry made mostly commercial aircraft flight management systems and instrumentation. They also had smaller space and military aircraft units. Their corporate culture was not terribly different from Honeywell's, with deep technical foundations. Most of Sperry was in Phoenix and I started taking a lot of trips there.

For S&RC it meant we gained two significant research groups, one developing flat panel displays for aircraft and the other researching fiber optic gyros. In Sperry, they had been separate groups attached to the related business units, unlike S&RC, which was a multi-tech central lab. Folding these groups into S&RC made the new ex-Sperry leadership very suspicious about how we would handle them. They especially hated government work because they didn't want anyone to see their books, and S&RC used many government contracts to supplement our research. Because of this, I made one of the wisest and most important decisions during my Research Director tenure. Whenever the opportunity arose, I encouraged all of our technical people to learn what Sperry needed and to start new technical developments to support their businesses. It would pay off big in 1991.

Health, or Lack Thereof

Most old people like to talk about their ailments, which is incredibly boring for younger people. Despite this, I'm going to describe some unusual health setbacks I've had along the way because they are just darn interesting, at least to me. They also provide a warning—do not let physical disorders derail a career. You can skip this part if you like.

This story began around 1985 when horrible pains hit my belly. I had just been promoted to Research Director, but by October the pains were unbearable, and in November my gall bladder was removed because of stones that had formed there. This was before the neat endoscopic, minimally invasive surgery was invented in 1987, and my operation resulted in a mean-looking 9-inch scar. But eventually I could sleep again—for a while. One nice side benefit was an amazing, 3-foot-long, get-well card from S&RC that I received in the hospital, which over a hundred people had signed.

Before long, a new constant dull pain returned. My spleen had grown to about five times its normal size. For the next 18 months, I underwent a series of increasingly scary tests to figure out why the pain wouldn't go away. The tests included a liver biopsy, tests for spleen cancer and leukemia, kidney tests, full-body MRIs, etc. During that summer, I occasionally snuck out to my car to sleep through the pain with Tylenol, or hid in unoccupied rooms in the new building. My secretary, Carol Warne, who had been with me since 1982, had a sixth sense about this and knew where to find me even when I didn't tell her.

Carol became a close friend and great collaborator during our years at

Honeywell, consistently handling all the details I missed. She understood that I was introverted and actively avoided new people when possible. Luckily, I was smart enough to know some meetings were essential, so I would just mention a need to Carol and soon the appropriate group would come to my office, having been scheduled through numerous calls and notes from Carol. She also helped keep the Center running during my medical emergencies.

In September 1987, the cause of my pain was finally discovered. One night I vomited about a quart of blood on the bathroom floor. In the hospital ICU later, about seven doctors were trying to figure out what had happened. Tubes went through a vein in my leg, fiber optic scanners were stuffed down my throat, and ultrasound scans and more X-rays were taken. My regular doctor, John Butler, was there when Dr. John Hughes made the diagnosis from my esophagus scans. A bleeding vein was sticking out of my esophagus wall, which Hughes fused shut with a chemical. Bleeding of that kind has a 30 to 50 percent mortality rate, so I am lucky to have survived.

Other tests confirmed that the many stomach and esophageal veins that were protruding were due to back pressure from a blocked portal vein. The portal vein is the main drainage line from all the intestines that feed the blood with food products into the liver, where toxins are filtered out. The vein can be about 1.5 cm across. If blood can't flow into the liver, the blood backs up, expanding the spleen and creating many new paths for the blood to flow backward into the regular system, mostly along the central digestive tract. At that time, no one understood why my portal vein was blocked, but that was the source of my pain. My body had essentially been re-wiring itself for over a year to handle its blood flow problem.

One side effect was a slow expansion of my abdomen. I later developed the look of a man with a beer belly without the compensation of a lot of beer intake. Heavy drinkers often get cirrhosis of the liver, which has a similar appearance. For the next 15 years I received regular esophageal scans, often with chemical sclerotherapy to shrink the protruding veins. Later, balloons were inserted as well to expand the diameter of my esophagus, which had shrunk due to all the scar tissue.

My skinny esophagus caused numerous embarrassing moments. Bread, hard rolls in particular, had a tendency to get stuck in the narrow spots. Often, when stressed or during important occasions, I forgot to chew carefully. At home, if I felt the choking feeling, I would drink water to force the food through, which usually worked. When it didn't work, it could be quite embarrassing, causing me to spit up or vomit uncontrollably. This happened once when I was on a stage, eating with other dignitaries in front of about 500 people. I made it out that time, but wasn't so lucky on another occasion. Honeywell had a big lunch in Washington D.C. with Dan Quayle, future U.S. Vice President, as the main speaker. During his talk, I felt a big bite of hard roll get stuck part way down. I tried the water technique, but it didn't work. When the water level reached the gag reflex area, I quickly exited the room and shut the doors just before barfing in the hotel corridor. After cleaning myself up, I returned to the dinner without mentioning the incident, until now.

A couple more episodes will complete my health picture. During the summer of 1991, I traveled to New York City for a two-day conference at the Waldorf Astoria. I had a bad headache the first evening, so I ordered room service to bring up a hamburger and tea. This was a little extravagant since the hamburger meal cost $40 with a tip. But the headache had gotten worse and I couldn't even eat my $40 hamburger.

Back home, the headaches worsened until I finally saw a neurologist. A brain scan revealed that another major vein was blocked that normally drained blood from the brain. Luckily, the brain has crossover points and multiple blood exits. Slow vein occlusions rarely cause permanent damage. But once again my body rewired its pathways to compensate for the blockage, a painful process.

I was told to use the blood thinner Coumadin the rest of my life. A blood specialist found the root cause. Apparently my bone marrow creates too many platelets, one of the key blood clotting factors. So I also needed to take a medication to slow down platelet production. Those precautions have prevented further clotting problems. I often wonder if a similar issue killed my Aunt Myrtle, who died at only 40 due to a clot in her leg that moved to her lungs—a pulmonary embolism.

Finally, looking at my MRIs, doctors noticed an odd formation in my left kidney, which seemed to be slowly growing. They decided it was cancer. So in October 1991, my kidney was removed. Fortunately, the cancer was well contained and didn't reappear.

A few weeks after the surgery, I rolled over in bed at home and heard a loud "snap" in my back. I woke Miriam and asked her to call a doctor so I wouldn't have to move. We finally got word back from the surgeon, who laughed and said that it was just a string holding my ribs together that finally dissolved. Apparently they needed a large hole to remove my kidney and they had pried my ribs apart. They apologized and said they should have warned me about the sound.

About a year later, a colleague from the Solid State Electronics Center (SSEC) developed a similar kidney cancer. I counseled him and visited him before surgery at the hospital, saying it wasn't too bad, but watch out for the rib snap sound. Unfortunately, his cancer wasn't contained and he died a few months later.

Recently, doctors discovered a similar spot in my right kidney, which we are watching very carefully. Losing that kidney would be bad, because you need at least one. But my current doctor isn't very worried because he claims that he could remove just a section around the small growth if necessary. Fortunately, the technology has improved during the last 24 years, and they rarely take out whole kidneys today. Too bad they didn't have that back in 1991.

Despite all my early health issues, I now mostly just worry about getting old.

The Girls

Through the 1980s, other than getting promoted and spending time in hospitals, I was adjusting to life with females at home. We spent a lot of time with my sister Janis and mom Fern, so I was outnumbered five to one.

Kelsey and Amy were good kids. Whether this was due to good genes, brilliant parenting, or blind luck, we don't know, but they rarely got into serious trouble. Miriam and I had very high expectations and Amy and Kelsey just sort of agreed.

Numerous activities filled their and our time: Yamaha piano lessons, dance and skating lessons, gymnastics, Kinderland and Pat's House (Pre-K daycare), Brownies and Girl Scouts, skiing lessons, kids' choirs, gardening, flute lessons (Amy) and trumpet (Kelsey), swimming lessons, T-ball, Gibb's

Farm (Pioneer School), a little golfing, and Shoreview summer camps. Each year I marked their progress with special birthday cakes, sculpted up to three layers high, with a design from their current obsession, like Strawberry Shortcake, Cabbage Patch Kids, shoes for running, math books, etc.

Each summer we went on trips to northern Minnesota, often with Fern or family friends. For two years we stayed at a place called Island View Lodge where the food was great and there was a show each night (sometimes with our kids participating). Amy caught her first fish there. We stayed at a Honeywell colleague's cabin twice, where I caught several large northern pikes and left them hanging on a dock-post overnight one time. By morning, crayfish had stripped them to the bone. The kids thought it was gross. I'm not much of a fisherman.

I always brought along painting materials for my leisure time by the shore. Later in the decade, we began staying at our friends' Barb and Chris Anderson's cabin on Bay Lake each year. They had daughters about the same age, Ingrid and Heidi. Amy and Kelsey learned to waterski there and we all boated to Church Island for Sunday services on the shore. Among many activities, Kelsey and Barb were always on a Pictionary team together and never won (Barb was a bit artistically challenged).

Both daughters were natural learners who pushed themselves. Amy was put in special programs for early readers as soon as she entered grade school. They created cool science projects. For example, Amy once measured and analyzed the force from magnets. Kelsey grew crystals one year and studied plant growth for another project. Both participated in something called Summer Academy for gifted students from many school districts.

Kelsey is particularly talented at drawing and art and focused in that area. Starting in 1989, Amy began five years of studying Swedish at Concordia Language Camp near Bemidji, Minnesota. She just thought Swedish would be fun and tied to her heritage.

Some of our most memorable times were on trips, and we took a lot of them. Here is a partial list:

- Oregon and northwest to Victoria, Canada ('82)
- Florida and Disney ('83 and '86)
- Boston and the Northeast ('86)

- Spain ('87)
- Philadelphia and New Jersey ('89)
- Phoenix ('90)
- Wisconsin and Door County ('90)
- Germany, Switzerland, and Austria ('91)

The 1986 trip to Florida, without Miriam, who was at home teaching, was bizarre. The first few days were freezing and we weren't prepared. Amy loaned a jacket to Kelsey, Janis a coat to Amy, Fern a sweatshirt to Janis, and I passed my heaviest jacket to Fern. As a result, I froze and developed a terrible cold.

Four-year-old Kelsey didn't understand the whole idea of Disney World but really liked the look of a pool at our hotel. So the first morning she refused to leave, demanding to be taken to the swimming area. The fact that it was freezing didn't register. So in the end we made believe we were leaving while she pouted under the sink cabinet. Going out the door, Amy said, "Don't worry, she'll come running after us in 10 seconds." Actually it took about a minute and we had started to worry.

Our trip to Spain was the most expensive "free trip" in history. Miriam had won an all-expenses-paid weeklong excursion to Spain as part of a Spanish teachers contest, where she had to fill out a questionnaire and write an essay about how the things she learned could be used in her classroom. Even better, the hotels for the trip were paradores, wonderful historic, remodeled buildings—the pride of Spain. Unfortunately for the kids and me, to tag along cost a fortune. Children were not allowed to stay in the same room as the parents, so we always had to buy an extra room.

We liked the idea of our kids seeing as much of the world as possible as they grew, so the cost of the Spain trip was okay. The first days in Madrid were scorching, averaging 115 degrees Fahrenheit during the afternoon. While the Spaniards were wisely taking their siestas, we roamed the city, sweating. After visiting the Plaza Mayor we trundled through the old town streets and up a steep hill, where we discovered a McDonald's. Amy pleaded for an ice cream cone and a chance to cool off in a familiar restaurant. To support her sister's plan, four-year-old Kelsey stopped in the middle of the street, hands on her hips, and said, "We're talking heat. We're talking feet. I'm not going one step"—pronounced "thep"—"further without ice"—pronounced "eith"—"cream." Later, revived, Kelsey went back to chasing pigeons around the city.

After Madrid we drove northwest aiming for Santiago de Compostela. At one lunch stop we ordered sangria to enjoy after the driving. But I cracked open the bottle at lunch and proceeded to get quite tipsy, setting a bad example for the girls. I teased Miriam as she drove, because she didn't dare pass anyone in the mountains. The kids and I screamed when she actually did try to pass. I have the whole episode on two hours of videotape. Amy and Kelsey also dove onto the floor from the backseat whenever they saw a bull on a billboard, yelling, "Bullboard, bullboard."

Our hotel in Santiago de Compostela was truly beautiful. The cathedral there is thought to be the final destination of the disciple St. John and is a major pilgrimage site in Europe for faithful Catholics. Our hotel, right next to the cathedral, was originally a hospital built by Ferdinand and Isabella in Columbus's time. Now thick glass modern doors separate the passageways while ornate carvings and gargoyles adorn the internal courtyards and walls. The combination of modern and ancient is stunning.

In the late evening, Spanish troubadours and college student singers (a.k.a. Tunas) roamed the small streets outside our open windows, adding a wonderful ambiance before sleep.

Back in Madrid, we saw four-year-old children roaming the city streets below our window at midnight. Amy and Kelsey thought they no longer needed to go to bed. On the final day we visited the Museo del Prado and El Retiro Park where we bought paintings and the kid got bright orange fans to cool themselves and show off back home.

My mom joined the girls and me on one General Management trip to Phoenix and Sedona. I would be trapped in several Honeywell meetings during our week stay, so Fern volunteered to come along and watch the girls so we could all get a winter break. In the evenings we spent tons of time playing miniature golf—there are great courses in Phoenix. Kelsey turned out to be excellent, once beating all of us with a long hole-in-one on the last shot.

The thing I most remember about that trip was an afternoon of legitimate golf at the McCormick Ranch course. We got a reduced rate as guests at the Scottsdale Resort, so we rented two carts. The plan was for Fern to drive one cart and me the other, with the girls renting clubs. The temperature was hot, so not many real golfers were on the course. I gave instructions to my mom about driving an electric golf cart and we were about to head out. But Mom's cart was set in reverse and she gunned the accelerator. The cart flew backwards and crashed into a large bush, with two actual golfers diving with their bags to avoid being crushed. After careful reinstruction, we headed out and slogged around for nine holes. I think Kelsey quit after a couple of holes. Water on that course claimed most of our golf balls.

Our girls seemed to grow up fast with many experiences. Some examples they remember:

1. Eight-year-old Amy got glasses and on the same day fell off her swing set and bent her arm bone to a 30-degree angle. A doctor put her under anesthetic and just bent it back to straight.

2. At the Tower-Soudan Mine they both went down 2,300 feet with hard hats to watch scientific experiments and see how miners worked a century earlier.

3. They swam on an Atlantic beach with our friends' children in New Jersey and attended the Mummers Parade in Philadelphia, blowing the big horns.

4. One day Kelsey went to second-grade class sick and barfed all over her desk and the little girl seated in front of her, who had just nicely offered to share her piece of cheese.

5. At 11, Amy ran in her first big race, something called the Turtle Kid Triathlon. She looked very cute.

By November 1989, the Berlin Wall had fallen. Miriam had kept us alive and together throughout these important years. I had learned to adjust to many females surrounding me, and loved them all.

Sports

Minnesota sports teams are almost always dismal. Four Super Bowl shots for the Vikings during the 1970s all ended in failure. One World Series versus the Dodgers in 1965—a loss. Timberwolves basketball and North Stars/Wild hockey—nothing.

There was one brief moment of glory Miriam and I will never forget. In 1987 we were forced to attend a dinner with my boss and his staff in downtown Minneapolis during the seventh World Series game between the Twins and the St. Louis Cardinals. The Twins had begun the season with a 1 in 150 chance of winning the World Series, which many thought was too optimistic. During our dinner we kept asking for game updates from the waiters since we could hear a TV set in the kitchen. We were the only table occupied in the Hyatt Hotel's most fancy restaurant. Eventually they brought the TV out on a stand and we watched the final innings, with the entire kitchen staff and waiters standing around the table. Frank "Sweet Music" Viola was pitching in front of a deafening crowd waving Homer Hankies. Viola actually lived very near my sister's house in Shoreview. We cheered every strike and out in the final inning, with the Twins leading 4–2. At the final out, everyone went berserk, at the stadium and at the restaurant. We stayed for dessert and watched the cheering fans out the windows walking back to their downtown condos.

The Twins won the World Series again in 1991, beating the Atlanta Braves, but the moments of sports happiness in Minnesota have been very sparse. There is always next year.

Vice President of the Systems and Research Center

In April 1988, I became one of Honeywell's vice presidents. There were about 60 at the time, one for about every 800 employees, but at that moment I was the youngest and one of the oddest.

Our previous boss, John Dehne, had moved on to head up the Electro-Optics Division in Boston, which made FLIRs, Forward Looking Infrared devices, mostly used on fighter aircraft and spy planes. The business was losing about $10 million per year due to problems making the product. John solved the crisis and stayed on in Boston a long time.

My responsibilities now extended to the "Systems" part of S&RC, which is where all the very large programs resided. For example, we led one of the three national VHSIC (Very High Speed Integrated Circuit) programs, which brought in over $50 million for new developments to Honeywell. Another big area was secure computing. Our lab essentially developed all the firewall and other techniques to protect valuable data for the government. That group was ultimately spun off as a successful new business. Several other big, multi-technology programs borrowed engineers from the research groups as needed.

For the first time I needed to understand and manage all the Center technologies. They included Sensors, Microelectronics, Control Systems, Signal Processing, Computer and Software Technology, and Human Factors. Speaking to and for scientists and engineers at the top of their field—over 50 percent had PhDs—was both challenging and exciting. The quickly evolving Computer Science and Software groups were the most difficult. I don't think they fully trusted people over 30 years old, and by that time I was 43, an old-fogey by their standards. However, I believe I did eventually understand most of their bytes and pieces.

Once again, getting the promotion wasn't easy, even though I was in line for the slot when I became Director. There was a new head technologist hired into the company, Len Weisberg, who was quite aggressive and wanted to look at external candidates as well. Len's title was VP of Research and Engineering. Along with other units, S&RC would report directly to him. The VP selection committee also included Roger Heinisch and Bill George. George had recently joined Honeywell after miraculously starting up the first microwave oven business at Litton.

Bill George is a rather famous executive, still featured in Harvard case studies, who later became CEO of Medtronic and a well-known business author. Anyway, after the selection process seemed to slow down, I wrote a letter to Heinisch, George, and Weisberg outlining my management philosophy in about two pages, together with my biography. The letter went through about ten drafts before I dared send it. There had been a large number of layoffs at the time in Honeywell, and I said I wanted to only hire the very best and pay them well. I must have said the right things because soon after the letter was delivered, my promotion was announced. Being VP of S&RC would stretch my introverted personality to its outer limits.

My sister Janis threw a very big surprise party at her townhouse to celebrate my promotion. All our relatives were invited as well as some full-size balloon people. She had great food, banners, everything. Kelsey, Amy, and I posed with the balloon people, and everyone had a great time.

The first thing I noticed at Honeywell once I became VP was that many people actually came to my meetings, even a few who hadn't been invited. They assumed I was talking for the corporation or had inside company knowledge to share. Occasionally I did have to officially speak to the entire S&RC, and the cafeteria would be filled to overflowing. At those meetings I talked about S&RC business or projects, and often about new company policy or organization changes.

Jim Renier had become the new CEO of Honeywell, which was nice because he was a founder of S&RC and a scientist, but he announced a new internal slogan: "Work smarter, not harder." Since I wasn't particularly into company slogans, nor were the highly educated people in the center, we pretty much ignored the new catchphrase, which caused some trouble because we didn't hang all the motivational signs and banners. We hurriedly put them up when Renier scheduled a visit. Another part of the campaign involved small containers with tiny rolls of paper that had some symbolic purpose, perhaps for jotting down our brainstorm ideas. Soon these were hanging on the walls of many of the scientists' offices with comments about toilet paper rolls printed nearby.

I definitely had to learn to speak to and for S&RC, and to listen carefully to everyone. I listened to great technical ideas while walking and talking to the scientists. I also listened to people talking about love triangles and restraining orders on our loading dock. I listened to employees talk about their emotional breakdowns or divorces. Somehow I resisted my natural temptation as an engineer to try to fix all the problems with suggestions. It turns out that just listening can be very appropriate.

Mostly, as VP, I tried to tell the truth, even if it didn't agree with company policy. For example, I once needed to explain that the company had decided to lay off about 10,000 employees. Everyone wanted to know how that would affect the Research Center. Our plan was to comply through attrition (i.e., not hiring for a while and letting the population drop naturally), but I refused to ape the company line about what a great idea the layoffs were. Wall Street had started to take over the management of companies by that time, meaning that short-term profits were all that seemed to matter to top management. Unfortunately, that attitude still prevails today, but hopefully it won't last forever.

The beauty of managing in a research center is that you never have to micromanage or lie. In large and small meetings, I simply tried to explain why situations were occurring or why certain inventions or behaviors would be useful. The key word was "WHY." Once the scientists and engineers understood why, they would do all the work and solve the problems on their own. Talking with them, I only needed to show genuine interest or concern about their work. They didn't need to be told to be smart. I am not sure if my management style would work in a big production business, but it was perfect for a research lab.

As my responsibilities grew, it was harder to keep track of the people and technical developments in so many groups and fields. Walking the halls and dropping in to chat with the engineers, scientists, and other staff could be a full-time job and not an efficient use of my time. I want to thank the dozen or more people who were my friends and key conduits to all the employees, people like Allen Cox and Howard French in the optics group; Brian Isle with the computer people; Fred Faxvog for gyros; Glen Sanders in the fiber gyro and Phoenix operations; Bob North, the human factors expert; Dennis Ferguson and David Lamb in the microelectronic area; Mary Hibbs-Brenner for VCSELs; Jim Lenz for all sensors; Tom Cunningham and John Weyrauch for controls; Anoop Mathur for home systems; and several others. These were the people I would have lunch with, who would drop by when something interesting happened, and who I could always tell the inside story and could trust to never lie. They were my scientific friends, and I couldn't have run a large research organization without their friendship.

One nice thing about being a VP is that you are also invited to many interesting external activities. Soon I became a member of the University of Minnesota's Institute of Technology (IT) Advisory Board where I met the leaders of technology in the state.

I also joined an organization called the Center for Development of Technological Leadership (CDTL), which Honeywell helped found at the University. CDTL was a kind of bridge between the Carlson Business School and IT. It taught subjects like strategic forecasting, leadership, marketing, finance, writing, and new product development. Renier had been a prime mover for CDTL because he believed that it's easier to teach technical people how to manage than it is to teach MBAs about high technology. Hundreds have graduated from what is now known as the Technology Leadership Institute (TLI). Based on the number of CDTL/TLI graduates who have become successful executives in many Minnesota companies, he was right.

One of Renier's early moves was to set up a structure that would one day create his successor. He named Michael Bonsignore and Larry Moore as co-Chief Operating Officers and Chris Steffen as Chief Financial Officer and let it be known that they were in competition.

I will never forget one personal occasion involving Jim Renier. The Aerospace Sector had what was known as a Technology Strategy Board (TSB), led by Len Weisberg at that time, which involved all the technical leaders, VPs of engineering, etc., from the business units. Responsibility for all the arrangements for one meeting in Minneapolis fell on me. James Renier was to be the guest speaker. So I contacted the Embassy Suites near the airport for the evening dinner. Here is a list of all that went wrong:

- The wine was bitter, and all the wine people in the group refused to drink it.
- The soup was too spicy to eat.
- The dessert profiteroles were completely stale. Profit was the major topic in the company at that time so I thought the choice was appropriate.
- Finally, Renier rose to speak. As a former scientist and the new CEO, everyone wanted to hear his thoughts. He began by telling us about a new company survey done by Human Resources that said the engineers across the company were unhappy. He then attacked all the engineering leaders at the meeting for not maintaining morale. He apparently never connected his laying off about 8,000 engineers as contributing to the morale problem.

Needless to say, I never went back to Embassy Suites or invited Renier to any future meetings. Now that I think about it, no one asked me to make dinner arrangements again.

Remembering my study about how competition brought out creativity and knowing how well the Initiatives Program worked, I soon introduced something known as the Home Run Programs. This was a competition among the most promising new product ideas at S&RC, with awards of $300,000 to $500,000 of development funds. I had attended a seminar at the Stanford Research Institute, where a methodology for evaluating the promise of technical ideas was developed. Their method involved estimating the cost, the likelihood of technical and product success, the business financial payoff, and the likely response of competitors. Eventually, I reduced their method to an Excel spreadsheet and a team evaluation process. By that time S&RC was receiving about $40 million per year in company funds, so we pulled about five percent from the science areas to fund the new Home Run Programs. After five years we found that the programs had a huge return on investment (ROI) for the company. In some cases, the ROI was over 200 percent per year once the businesses got going.

One example was a new ring laser gyro for the aircraft navigation business. The related business units thought it was a dumb program because the existing RLGs were selling just fine. So we funded the effort using Initiative and Home Run funds for five years until the device worked and customers started wanting a new device. At a meeting with the business people, we put the gyro down on the table and said it worked better than the existing gyro, was cheaper, and smaller. Suddenly, they wanted the gyro and told us that they no longer needed S&RC's help. That's one trouble with being in R&D; you don't always get to stay with your inventions and see them mature. The small RLG, the 1320, is now a major product line for Honeywell.

Honeywell's annual General Managers Meeting was usually held at some swanky resort. When the financial results for the year were good, the parties were first-rate. Spouses were invited, but Miriam couldn't always attend. At evening get togethers with lots of drinking, I would huddle with a few fellow introverts while many of the executives were making a lot of noise. Of course, business strategy for the coming year was also discussed during the day.

At one such meeting we noticed that among the 60 or so VPs, only three had spouses that still had jobs, and that included the four or five female executives. The three were Miriam, Larry Welliver's (Head of the Solid State Electronic Center) wife, and Steve Hirshfeld's (our friend and head of company strategic planning) wife. Later, both Larry and Steve's wives would also quit their jobs, so Miriam was the only holdout.

Once, during an afternoon session with all the VPs, a consultant asked what each of us would do with an unexpected gift of $100,000 from the company. We went around in a large circle, each explaining what they would do. Every single executive there said they would buy something, except me. I said I would save the money and invest it. Perhaps that shows how unfit I was to be among the executives. It also shows what Miriam and I were doing with money. We are both rather frugal (Miriam bizarrely so). I occasionally enjoy buying a new car or computer but not much else, while Miriam loves finding a bargain at a garage sale. Back then we were making and saving a lot of money, which was being invested fairly aggressively and successfully by a financial company.

Let me relate one final story about my time as Vice President of S&RC, and perhaps the shrewdest management move I ever made. Generally, I hated the Monday morning staff planning meetings for the week because everyone felt the need to talk about their routine work and the meeting dragged on and on. So I sometimes substituted brainstorming sessions or long-range-planning meetings. One of those meetings was very beneficial.

That day, as a team, we speculated about what would happen if Honeywell sold off its military business. One quarter of S&RCs technical developments were in support of military businesses and about one fourth of our funding. After buying Sperry Aerospace in 1986, a non-military business, Honeywell seemed to be leaning away from the bombs, bullets, and torpedo businesses. So the Center's leadership team developed an entire plan on the hunch that those businesses might be sold one day.

Since the late 1960s, an organization called the Honeywell Project had been periodically picketing military products outside Honeywell's gates, both before and after the Vietnam War. The campaign reignited in the 1980s when a campaign against land mines and cluster bombs became intense. Honeywell's top management simply became tired of the continued bad exposure and, in September 1990, they spun off the Defense and Marine businesses by creating a new company, Alliant Techsystems.

A couple weeks after the internal discussion of this move started, several defense businesses' top officials visited S&RC, probably to say they wouldn't be supporting the Center anymore and that our internal funds would be dropping drastically. We were able to pull out our secret "what-if" plan, which involved transferring S&RC people directly involved with their business to the new company. When they saw our logic they tentatively agreed. Len Weisberg was astounded that we had a fully developed plan ready only a couple weeks after the news. He started referring to me as a genius.

Of course that wasn't the end of the discussion. When Roger Heinisch and I announced to the affected employees that they were being transferred, they hated it. They preferred being in a research center and had no idea what their fate would be in the new company. We were transferring the entire Signal Processing Science Area and many individuals from other groups (about 100 in all). This was the most difficult meeting of my life. No matter how much we explained the logic and why they would be better off, the situation was very emotional for those involved. Even though the transfer eventually went through and a new research group was established at Alliant Techsystems here in Minnesota, it was very difficult.

One of the transferees was Mel Geokezas, my one-time Co-Director. Tom Cunningham, from the Controls Group, was selected as the new Director. Also soon afterward, I received the new title of Vice President of the S&RC *and* the Solid State Electronics Center (SSEC). I now had gained a $100 million troubled production business to run, as well as the S&RC.

The 1990s

Entering the 1990s, our family enjoyed a normal, if somewhat quirky and hectic, home life with the girls growing up fast. More stories about them will follow later. We had our old deck removed in 1989, replaced with a four-season porch and a two-level deck with a storage shed tucked underneath. The cost was about $35,000, (it would probably be about $70,000 today). I was through doing major projects myself, so we hired Rossbach Construction to do the work. The four-season porch became a true joy. In winter I would sit out there in shorts and a T-shirt with the temperature cranked up and the ceiling fan circling slowly, looking at the snow all around and imagining it as white sand on a tropical beach.

We took some fine trips around that time, to Phoenix; Door County,

Wisconsin; and to Europe. The Europe trip reprised one taken by Miriam and me in 1975, to Switzerland, Germany, and Austria. The girls did a *Sound of Music*-like spin on a mountainside in Austria. Later, on a hill overlooking Vienna, we stumbled onto what seemed like an old Nazi outdoor restaurant. It looked very clubby, everyone drinking beer, but it had a great view of the city below. We arrived about 4 p.m. and it took about two hours to get and eat our main food. The waiter kept ignoring us and looking away whenever we waved for service. They must have hated children or Americans. Finally, we ordered ice cream desserts and patiently waited for our bill. The bill never came, and the sun was setting. So I estimated the bill and we all got up and walked away. The waiter finally noticed us and chased us down in the parking lot. I said we left money on the table and got into the car. Other than that, it was a great vacation. From the top of a hill in that area, we looked deep into Czechoslovakia but couldn't visit due to my top-secret clearance and because I hadn't prearranged such an excursion.

In the 1990s, my impression was that older Europeans were still very much into hating other ethnic groups. After Honeywell diversity sensitivity training, of all things, several German executives left the room making jokes about black people. That evening, after drinks with those managers, they talked judgmentally about the differences in people, even in small subregions of countries, and all their historical offenses. I wasn't glib or courageous enough to argue, which made me feel guilty and sad afterward. Hopefully young Europeans are now better.

The early 1990s were a difficult time for Honeywell's aerospace businesses. After the Berlin Wall fell and the USSR sort of collapsed, many government contracts were drying up. My first years of VP-hood dealt mostly with contraction. We did stop hiring new scientists and a few of the managers were quietly told to start looking for new work. Rarely did we force anyone to leave but some were encouraged. This was an issue both for S&RC and my new business, the Solid State Electronics Center.

SSEC was located in Plymouth, Minnesota, off of Highway 55. They had excellent facilities for producing semiconductor products—clean rooms, large vacuum stations, automated disc-handling equipment, etc. They also had a strong leader, Larry Welliver, who truly understood the technology well. Their main product line was radiation-hard electronics, devices that were used in most U.S. satellites, commercial and military, to prevent ionizing radiation from the sun from damaging the electronics. Like the rest of the electronics industry, they were in a never-ending battle to make smaller and smaller devices. Packing more transistors on a chip required ever newer and more expensive equipment, up to tens of millions for a new line. So my job with SSEC was primarily to oversee their equipment expansion and business lines. They gave me a nice office in their facility and I tried to spend at least one or two days there each week learning their business and the technology.

They also had an important business making pressure sensors for Honeywell's Industrial Automation business, which was slowly contracting at the time.

My main contribution was to persuade them to start new product lines to balance their revenue sources. For example, I encouraged SSEC to produce a line of tiny magnetic devices, which later led to many sales to General Motors. The little direction letters you see on rearview mirrors involved these magnetic sensors. They also began to make magnetic non-volatile memory chips that could store data even when the power supply was removed.

I really didn't do much to change their business, and soon the satellite chip business strengthened and I let them grow on their own. There were several strong leaders in that business, and benign neglect seemed like the best approach for me. Later, they took my office away because they needed the space for meetings.

During those days, I always had a ten-year plan that I discussed only with my closest friends. The plans often included ideas that were clearly impossible to act on quickly. One of those dreams was to unite the research across Honeywell to create more flexibility for the scientists and engineers. Starting with S&RC, I had always felt restrained working only for the Aerospace businesses. I still had misgivings about working on defense projects, even if they were never offensive-weapon related. The government was also moving away from developing military-only technology. Even DARPA, the Defense Advanced Research Project Agency, was demanding that our developments also have regular commercial applications to keep the cost down. DARPA was developing the internet, as an example, at that time.

From time to time, events would occur that made it possible to take small steps toward multiple ideas on my ten-year plans. When that happened, I could, as a VP, move quickly. As an example, when management was complaining about the cost of maintaining Honeywell's Corporate Research Center in Bloomington, I started talking quietly about that Center possibly moving in with S&RC and SSEC. Those two buildings, with all our attrition, now had enough space and facilities to accommodate the corporate research people. Ultimately, I convinced Roger Feulner, their VP, to make the move. Even though the organizations weren't united, it would now be possible for people to easily work across former barriers. Also, the old solar energy group moved in with us, bringing back one of my first research loves.

I believe this is a very important management and development strategy: prepare for the major technical or management trends you want for the future and then when an event occurs allowing you to move ahead on multiple fronts, pounce on it.

One huge opportunity came to fruition in 1991. Honeywell's Commercial Aviation business won the development contract from Boeing for the new 777 aircraft. Dubbed the "zero-maintenance" aircraft, it had been the big target the aviation people had been going after throughout the 1980s. Of course it still required maintenance, but that was greatly reduced by Honeywell products. And S&RC had major contributions to this win. For four years we had asked every group in the Center to do whatever they could to support this former-Sperry business line, and did we ever. About 14 major new inventions developed at S&RC went onto that plane. Some examples:

- A central redundant computer and secure software, replacing dozens of small computers throughout the plane. Previously, every small sensor or actuator needed a dedicated computer to function. We brought those all together. Keeping routine software from interfering with flight-critical functions was a major breakthrough.
- A new RLG (the 1320) inertial navigation system.
- A new fiber-optic gyro system.
- Flat panel LCD displays in the cockpit for both the pilot and co-pilot. They were the first wide-angle viewable displays, so both pilots could see both displays in case one went out. This technology quickly spread throughout all display manufacturers in Japan and

made possible the wide-angle view on today's LCD TV and computer screens.

- A central maintenance system.

There were many other S&RC developments that didn't get on the plane. Boeing's procurement people believed they were becoming too dependent on Honeywell with this contract. My pal, Jim Lenz, had developed a new sensor for the doors and wing flaps to determine when they were fully closed, and he didn't like the idea that the sensor hadn't been put under development. So he tried an "end-around." He showed his sensor to the Boeing engineers he knew and got them to demand that it be put on the plane, which caused a great deal of embarrassment for the President of Honeywell's Commercial Aviation business who called Jim and shouted at him. Jim, in his normal fashion, shouted back and got into big trouble for a while. But overall, the 777 win was a tremendous success for Honeywell and S&RC. Honeywell's normal portion of the cockpit avionics rose from about 30 percent to 65 percent on that contract. The technology went into production on the 777 in 1995 and was later extended to the 767, 787, and all new versions of the 737. Many of the Honeywell products are also on Airbus jets. That contract has led to sales of at least $11 billion so far.

The early 1990s were a time of great activity both for our family and my work. Before returning to the story of Miriam and the children, I need to describe three amazing trips I made for Honeywell.

The first was with Orv Bratlund, my new Director of Business Development. Several technology-sharing opportunities had appeared in France related to fiber optic sensors. Before going to Paris, we were to meet with the President of Honeywell Europe in Brussels to be briefed and shown around. The trip went wrong from the start and we arrived at Schiphol Airport in Amsterdam very late. Orv's luggage had been lost, and he kept saying that he refused to meet the President of European Operations without his suit. The next morning his luggage arrived, thank goodness.

We found our car, an Audi, in a rental parking lot but couldn't figure how to get it into reverse. I know this sounds stupid, but even after finding the owner's manual and trying to read it in German, we couldn't get the car to move. Finally giving up, we stood out beside the car until a Dutch man walked by and saw our distress. He seemed to know about the secret small button on the side of the steering column that allows one to get the car out of park. We then drove to Brussels.

The Honeywell people there told us about some opportunities and we briefed them on our technology, and then off we flew to Paris. Honeywell had one salesperson in France, an interesting character who saw dozens of business opportunities for us, most of which were lame. We did have a wonderful hotel directly across the street from the Arc de Triomphe and a little free time to visit the Sacré-Coeur church. I learned how to order a croissant and coffee the next morning at a small restaurant, then off we drove to the target company, whose name I can't remember. The company mostly wanted to let us know about some patents we could buy from them and especially wanted to jointly develop fiber optic gyros. But the real highlight was the lunch in a room off the company cafeteria. Apparently seven-course lunches were the norm for guests. We had a cheese and fruit starter, hors d'oeuvres, and several more plates. The whole thing ended with ornate boxes of cigars being passed around, which I declined, but everyone else lit up. Perhaps they thought a smoke-filled room would soften us up. We left for home with no official deals but some new friends we would bump into occasionally.

The next trip of note was the Jim Renier "Swan Song" trip. Our CEO was getting close to retirement and needed an excuse to visit all his favorite organizations in Europe one more time. So he announced a "technology assessment" tour, where he would nominally ensure that the scientific future of the company was safe. Honeywell had factories in almost all major European countries. Len Weisberg, Roger Feulner, and I were invited along on the Gulfstream IV to help him evaluate. The Gulfstream had fold-down seats that functioned as beds, which was nice except that Len's stocking feet kept touching mine. That trip is where I first learned how CEOs travel.

We had one dinner on a boat on a Dutch canal and later a ride on a Rhine cruise ship. The hotels were amazing—beyond five star. In Bonn we stayed at the hotel that heads of state frequented, with glass elevators and strict security. In Paris, after technology tours and a long seven-course lunch in the cafeteria, we prepared for dinner. We rode limos onto a large estate and pulled up to the entrance of large, white, chateau restaurant. Lights and candles were burning everywhere. Honestly, I can't remember anything about the food, but the ambience was unforgettable. We had the entire two-story marble chateau to ourselves, a long table seating 40 people from Honeywell, with a live violin group playing in the background.

After visiting Spain we flew to Italy for the grand finale in Milan. I recall that the rooms were $600 each, or about $1,100 today. My room had inlaid wood walls and ceilings and antique furniture. The bathroom was half as large as our living room at home and all marble, even the ceiling. I doubt that the Renier swan song had any lasting effect on the business, but it taught me a thing or two about extravagance.

The whole subject of executive perks is provocative. They become a wasteful way of life for many near the top and are designed to create jealousy among those not quite up the corporate ladder yet. Clearly many people exploit the opportunity, indulge in excess, and show off. For me, I admit I loved the new car I received every third year. But I turned down other benefits, like a free membership at a golf resort or business club, and the Honeywell home security system.

Of course the greatest incentive was money. Toward the end of my career, over 50 percent of my total compensation each year came from bonuses, restricted stock, or stock options. The percentage is much higher for those at the very top. Incentives undoubtedly drove decision-making and possibly work intensity for some, although most worked harder than was healthy already.

My main criticism is that what was measured and incentivized was wrong. Our compensation was determined almost entirely by short-term financial and stock performance. If I were in charge, the priorities would have been (1) customers, (2) employees, (3) owners, and (4) community. I think the stock market would take care of itself if the main concerns were in that order.

Japan

My trips to Japan were so interesting that they demand a section of their own. The first was in 1984; an S&RC manager named George Bynum set up the trip and accompanied me. Yamataki-Honeywell, our Japanese affiliate, had created a research operation headed by a man named Kozi Tanaka, who would become my lifelong friend. I remember flying in alone and asking the taxi driver to take me to "Hotel O-kur-a," praying that I had the pronunciation correct. I must have been close enough; I soon arrived at a beautiful hotel with nine restaurants, only a few blocks from the Imperial Palace.

My first impression of Tokyo was how three-dimensional it seemed. There were crosswalks at the five-story level, three-level freeways over the rivers, and the underground was alive with trains and people.

The Yamataki people treated us royally. We visited the electric center of town, Akihabara, like Times Square only more so. I mentioned that I would like to visit a department store, so our Mercedes chauffeur drove around the building for a half hour while we shopped inside.

I am not a huge fan of Japanese food, but we had a wonderful dinner in a Yokohama restaurant that we had to find on the subway by ourselves. Luckily, I quickly picked up on the Japanese Kanji symbols and we got off at the right station. We ate all evening, with beer, whiskey, and sake to help. I developed a strategy of just asking what foods I was not supposed to eat, after starting to bite a flower and having people say, "No, no, no." I loved the tempura and the cooking of Kobe beef in a hot oil container on the table, and Kozi Tanaka was a great host.

Of course we visited the Yamataki factories, where we saw the morning exercises everyone did together and learned about the new concept of "quality" that they were using and that would later spread throughout the world. They wanted to emulate S&RC in their new research operation so we told them our general approach, which began a long period of having young engineers from Yamataki coming to S&RC as researchers and to learn about our techniques. One young engineer was Ken (Takeshi) Kawai, who took us to the temple and shrine at Kamakura and generally showed us around. He instructed us on how to eat at a sushi restaurant where the food came on a conveyor belt and also took us to the Pachinko gambling rooms. Ken would be our first visiting researcher at S&RC and, surprisingly, turned up in 2014 with a request to be described later.

A second Japan trip in 1991 was amazing for many reasons. I sometimes call it the "Trip Where Honeywell Sued Japan." Traveling with my colleague Jim Lenz, the main purpose was to connect with Hosiden, the company selected to manufacture Honeywell's flat panel aircraft displays that our Phoenix research operation had been developing. The panels would have the new wide-angle viewability and we wanted to see the large vacuum process chambers, make sure they were adequate, and meet the people responsible. We flew to Tokyo and then to Osaka, staying in a fifty-story building overlooking the city. After driving to Kobe and inspecting the factory, our hosts planned to take us out for a special dinner. This turned out to be a little confusing for me. Each person at the table had a "female assistant" dressed in traditional garb. Our "assistants" cut our meat, filled our glasses with alcohol, and generally nudged against us. Later we were encouraged to dance with our assistant, and still later, approaching midnight, sing karaoke songs with them. I started to wonder how far this would go and suspected Jim Lenz had set this all up. Fortunately, we all just went back to our hotel exhausted.

Jim and I tried to take the bullet train back to Tokyo to visit the Yamataki-Honeywell Advance Center and our friends there. But getting on the train was no cinch. There were no obvious places to put large luggage near our seats. So we moved several cars away, with people staring at us, to find a small nook to stuff them in, and then worked our way back to our seats, hoping the luggage wouldn't get lost. We passed Mount Fuji on the 200-mile trip and watched what seemed to be a continual city fly by while traveling at the 10-story level part of the way. We saw people sitting on the roofs of tall apartments, looking down on them as we traveled at about 180 miles per hour.

Finally in Tokyo, we met with Ido-san, the CEO of Yamataki-Honeywell, in a conference room and then walked together downstairs to an indoor parking lot to drive together to a golf course. Throughout, the actions of female support people surprised me. They came to the conference room with tea and snacks, bowing before carefully backing out. And most unusual, one ran ahead as we moved toward the cars, opening hall doors along the way and, finally, the car doors.

The hilltop golf course was beautiful, known as Fleur, "flower" in English. The fees were probably astronomical, but Ido-san was a member there. Now, as I've admitted, my golf skills are minimal, and that was demonstrated for all to see at Fleur. Jim Lenz, Tanaka-san, Ido-san, and I started off with two women caddies wearing wide-brim hats and veils. They chased down our golf balls, even when I knocked them into thick forests lining the course, always finding the ball. There were two greens, winter and summer, for each hole. But the part I remember most was hitting my ball into one deep sand trap. The ladies shouted, "Boonker, boonker," to let me know I was in big trouble. These traps were very deep, and I bounced a few shots off the lip, each one rolling back into the trap. Rather than giving in, I kept hacking away, about 15 times, before giving up and hitting away from the green to get a clear shot. Ido and Kozi thought it amusing, since they were both great golfers.

On the final day in Tokyo, we were scheduled to meet with Canon to talk about technology sharing. But around 5 a.m., Kozi got a phone call telling him to cancel the trip to Canon. While we were at the research office, the news broke that Honeywell had sued about a dozen Japanese companies and hadn't bothered to tell anyone at Yamataki-Honeywell about the action. Every famous camera or camcorder maker was on the list. They were being sued for stealing the technology for the auto-focus. Developed in the 1970s by a Honeywell Corporate Fellow named Norm Stauffer, the technology had been shown for a $50,000 fee to numerous camera makers. By the early 1980s, most had incorporated the device into their cameras without compensating Honeywell. So in 1991, Honeywell's CFO, Chris Steffen, decided to sue them all. Chris Steffen was one of the three people competing for the CEO position when Jim Renier retired and probably wanted to make a final big splash. He had been a driver behind the sale of Honeywell's defense businesses and the main reason thousands of people had been laid-off. To give you a sense of his insensitivity, he had once been driven to the Minneapolis military factory in his Rolls Royce just after massive layoffs there. He was an arrogant loner who once complained that no store in Minnesota could sell him the $5,000 mattress he was accustomed to, so he had to fly to Michigan on the corporate jet with his family to go shopping.

Anyway, the news that Honeywell was suing several Japanese companies did not play well in Tokyo. Ido-san was dragged in front of the press and TV people and asked to explain—a major face-losing event. Jim Lenz and I spent the day apologizing for the lack of respect shown to our friends. Basically, a 40-year successful arrangement between the companies was destroyed. A few years later, Yamataki changed their name and severed the once-close relationship.

It's not that the lawsuit was unjustified. In the first actual trial against Minolta, engineering notebooks were displayed showing planes dropping bombs on a sketch of the U.S., saying they would ruin our camera business using our own technology, repaying the U.S. for World War II. This was the smoking gun that led to over $120 million in payments to Honeywell from Minolta. Later, the other companies complied without trial and royalty payments to Honeywell of over $600 million were collected. But I will never forget the sadness among the regular troops at Yamataki's research operation the day the lawsuits were filed before we took the long flight back home.

Our Family

I certainly didn't have a nine-to-five job and was often out-of-town for work, but in many ways we were a close family. The girls say they remember fondly our morning breakfasts together after Miriam left early for Como Park High School. The kids and I often popped in Eggos or shared cereal in the morning and talked about school. Most evenings we all ate supper together at the kitchen table, with me cooking on weekends and Miriam the rest of the week.

They were smart kids and did very well in school. Amy started taking math classes at the University of Minnesota in a program called UMTYMP (University of Minnesota Talented Youth Mathematics Program) in 1992. She completed four years of high school math and then three years of college-level calculus during the five-year program. She also met many students from throughout the Twin Cities area who became lifelong friends. In 1993 Amy came in second in something called Math Counts, a metro-wide contest.

Kelsey, who also did well in school, became our family athlete. She played midfield in soccer for many years because she could run forever. During winter, she played on local basketball teams. One year, Kevin McHale was assistant coach of her middle school team (his daughter was also on the team). Kevin, an all-NBA player for the Celtics, later became coach of the Timberwolves and Houston Rockets. He tried diligently to teach the pick-and-roll to the young girls, with modest success.

We especially enjoyed traveling together as a family. In June 1992, we drove to Champaign, Illinois, for a reunion among all of A. C. Anderson's graduate students. As the oldest among those attending, I had to give a dinner speech. It's a testament to Andy and Janet Anderson's warmth toward their students that so many returned for his reunion.

I learned that a mistake had been discovered in my thesis research. On one of many metals tested, an experimental error had matched a computation error in my theory. So the theory still looked good, but I'd made an embarrassing minor slip-up.

The kids enjoyed playing softball at a park, seeing the campus, and also my intricate refrigerator plumbing in Anderson's lab. We also made fun stops to visit Steve Landy and family in Indianapolis and Miriam's aunt and uncle, Adrienne and Albert Cox, near Chicago.

The family also took an interesting cruise in the Caribbean in November 1992. Starting in San Juan, Puerto Rico, all went well until we cleared the east end of the island. Hitting heavy winds, the ship began to heave, and so did many passengers. Only a third of the travelers showed up for the first fancy dinner on board. We were there but couldn't eat much, and the staff was handing out Dramamine pills with the food. Our children were not happy with boat travel, but the wind died down and the next day we arrived safely in St. Thomas.

There was a banana boat strike on the island of Guadeloupe, so the kids and Miriam joined a Bingo game the second evening while I wandered around the boat. Ten-year-old Kelsey won about $100 that night. Amy won a smaller amount and I worried that we had planted the seeds for lifelong gambling.

Basically, Kelsey is a lucky person. Winning seems normal for her, despite all the rules of probability I could cite. The second evening, surrounded by her posse of young friends, who annoyed the adults by winning too often, Kelsey hit the jackpot. Bursting into our cabin, she yelled, "I won, Daddy!" while waving a large wad of cash in my face. We counted it together—over $800. She had won the grand finale, shocking the large crowd. Miriam forced her to pay back the bingo entrance fees for the family and save half. Kelsey spent the rest of the trip buying nice gifts for us. Kelsey and Amy started to think cruise ships were okay.

The rest of the trip took us to Grenada, Venezuela, and Aruba. Coming into Venezuela in late evening was beautiful, with stars shining overhead and lights all along the northern coast of South America. We saw the sky-scrapers, government buildings, and slums of Caracas the next day before our bus climbed the mountains to a famous glass-blowing factory. The very next day gun fighting broke out at the government buildings we had just vis-ited as a coup attempt began. From jail, Hugo Chavez had engineered an uprising that would bring him to power a few years later. We were lucky to miss the gunfire.

Probably the best trip our family ever took was to England and Scotland in 1993. A local England specialist custom designed the lodging for the trip, taking into account family details. We had a great flat in the upscale Kensington area for our first week in London. The girls, ages 10 and 14 at the time, got hats in Covent Gardens and began wearing them everywhere. Soon Kelsey began successfully imitating the English accent. The royal fam-ily was having some financial problems and for the first time Buckingham Palace was open for tourists. We saw all our favorite sites and attended *Les Misérables* with the original British cast.

We played a CD of the play's music almost repeatedly as we headed north in a green Jetta rental car—the words now permanently seared into our brains. But all was not well with our Jetta. Flies started appearing, which we swatted out the window at first. But they kept coming out of the air vents, hundreds, we were counting. Heading toward Kent, swatting flies and sing-ing *Les Mis*—what fun. Finally, one morning, we found the car seats and floor covered with dead flies. Eggs must have been laid somewhere in the ventilation system, and we had to have the whole car fumigated. We had a forceful word or two for the rental company upon returning to London at the end of our trip.

But to show you how interesting our lodgings were, consider our stop in Scotland, near Edinburgh. A family with two young girls, similar in age to Amy and Kelsey, lived with the mom and dad and several dogs in a 1,000-year-old castle they were restoring. The girls became instant friends. These were more than simple bread and breakfast stops; they were family-to-family encounters only offered a few times each year.

After playing an embarrassing round of golf and visiting the Loch Ness monster, we stayed at a series of fascinating homes throughout England, including:

- A large estate home with a master bedroom half as large as our house and a wall of windows overlooking rolling grass fields (like Downton Abbey). Kelsey said she felt like a princess.
- A townhouse, where the 30-something son quizzed Amy relentlessly at dinner on all her beliefs, using the Socratic methods he'd learned at Oxford.
- An amazing 500-year-old house near Stratford, similar to Shakespeare's home, with low-beamed ceilings and tiny doors perfect for people shorter than 5-feet tall. I smashed my head getting up in the night—twice.

We saw Stonehenge, the Lake District, Bath, Canterbury, Dover, and took a hovercraft to France for a short stop. That month-long vacation created wonderful shared memories for us all and makes me want to take our grandchildren there someday. Kelsey later chose to study in Cambridge for a semester because of that great trip.

VP of Technology

In April 1993 the dominos began to fall. Jim Renier had finally retired as CEO and Michael Bonsignore was chosen as his replacement, rather than Larry Moore or Chris Steffen. Moore became Chief Operating Officer, now leading all business operations and Steffen left in a huff. Good riddance. Len Weisberg also retired at about that time and I, miraculously, was promoted to Vice President of Technology for the entire corporation.

Bonsignore's selection was not too surprising since he was an engineer who led the Marine Operations before heading up all of Honeywell Europe, so he had the most international and general Honeywell knowledge. Larry Moore had come from the Sperry Aerospace acquisition and thus didn't have long Honeywell credentials. But after this change, I reported directly to Larry Moore, someone who greatly appreciated all the research support we had given to his Aerospace divisions and whom I greatly respected.

Bonsignore was a smart schmoozer, more of a spokesman and salesman, while Larry Moore cared deeply about products and understood how hard they were to actually create. Larry also promoted a few other quiet people, like Arnie Weimerskirch, to head the Quality Programs, people who knew their stuff and could be trusted.

Here is a rough picture of Honeywell's main businesses at the time, without much detail because reorganization of the subunits occurred regularly. The three main business lines were where the money was made and were stable through the 1990s despite occasional acquisitions and divestitures. The international units sold both U.S. and locally produced items.

About 40 percent of the business was Home and Buildings, 30 percent Industrial, and 30 percent Aerospace, although the highest profitability was in Aerospace. The defining word for Honeywell at this time was "Controls."

AEROSPACE	Aircraft flight management, displays, commercial and military navigation, and gyros. Satellite systems.
INDUSTRIAL AUTOMATION	Refinery flow and pressure sensors, system and safety process. Factory process management and products.
HOME AND BUILDINGS	Home thermostats and security. Building heating and air-conditioning products and systems.
NON-U.S.	A myriad of similar products and systems.
EVERYTHING ELSE	Finance. Legal. Human Resources. Quality. Research and development. Specialty electronics.

Early on, Moore let me know that my job wasn't just to run the research labs. He was worried about the future of the Industrial Automation business in Phoenix, which produced the systems and sensors that ran large refineries and factories. These businesses had always been managed by our huge, expensive systems like the TDC 2000 or 3000, which sold for millions each. Developing a new system could take five years, but technology was moving much faster than that. So Moore told me to pick the 20 best engineers in the company and do a total assessment of the Industrial Systems business. Fortunately, he had warned the new Vice President of that business, Markos Tambakeras, who agreed with the need for a fresh viewpoint.

After about six months of work, my team suggested major changes to the product line and the development process:

- Take advantage of the emerging web system technology, but keep it away from the critical refinery processes.
- Create an iterative software process, rather than the usual waterfall requirements approach, to speed up development.
- Make the software and hardware more modular, so future improvements could be done on the fly.
- Put more of the product value into the software rather than hardware.

Finally, we evaluated the capability of the existing management team. A dozen engineers verbally volunteered that the existing managers were not capable of leading them to the next generation of products.

We gave our report and got out of there. Some people got fired, but a couple years later they came out with a totally improved approach to their business and a new Process Solutions line that became the most successful new product in Honeywell history with over $50 million in sales the first year.

Leading Red Teams to evaluate or improve company products became an important part of my job. These temporary teams often borrowed leaders from throughout the company to solve current major problems or explore future improvements. The division general managers never appreciated lending key personnel for my teams, but I had complete backing from Larry Moore and most of the studies were his suggestions. Here are some subjects of the Red Teams: (1) wireless products for the Home & Buildings and Industrial businesses, (2) use of common systems architectures, (3) a survey of possible future sensors, (4) security in Honeywell products, (5) faster product development techniques, (6) intellectual property strategy, (7) a new home-systems vision, and (8) solving production problems with the important gyro/inertial measurement units.

I now had the power to start implementing some of my long-range goals, but I never just demanded a change. For example, I persuaded Roger Feulner and his Corporate Research Center to simply merge with S&RC, forming a new entity, the Honeywell Technology Center (HTC). At last, S&RC people could work directly on projects in the industrial, buildings, and residential markets. A new VP was needed for HTC and a person from the non-S&RC groups, Ben Simmons, was chosen. This helped build good will between the groups. All the science areas became stronger due to the changes. Roger Feulner took on a new challenge, to form a software operation in India, which will be described later.

At that time, five people with the title of Vice President reported to me as well as one who was nominally a President. The President was Henri Hodera, leader of a small group left over from a much larger ($80 million) Honeywell acquisition called Tetra Tech, which Henri had founded.

Tetra Tech made underwater robots, fiber optic devices, and other products. After spinning off most of Tetra Tech, Henri and his small group in San Diego remained doing classified government work. His group refused to report to one of Honeywell's business units but liked working for HTC, because we, at least, could understand what they were doing. Henri is an amazing person and friend, still leading new companies and doing important research at age 88.

While most of Honeywell's 10,000 engineers worked officially for the business units, my small team monitored the development processes they used. Gary Kirchner, and later Tom Fletcher, had the title of Director of Engineering Process and could impose goals for the divisions. The single most valuable goal implemented was a measure for the rate of new product introduction. This mirrored a 3M Company's strategy. The goal was to derive 30 percent of all company sales from products introduced in the prior three years. When we started measuring in 1994, about 12 percent of Honeywell's total sales of $6 billion, or $700 million, were from new products. By 1999, over 31 percent of Honeywell's $9 billion in sales, or $2.79 billion, were from new products. The late 1990s were very good years for business in general, but I am proud of the role our new products played in Honeywell's rapid overall growth (nine percent per year) during that period.

So beginning in 1993, Michael Bonsignore often referred to me as his Chief Technology Officer. Carol Warne was my secretary again, one of about eight people on my direct staff. We had an expenses and travel annual budget of over $2 million. The staff evolved over my years as VP of Technology, but the team was very strong and enabled us to accomplish many goals. Early on, George Bynum was the group sage (I can't remember his exact title) but he had worked at Honeywell for many years and shared his wisdom, especially with me. Orv Bratlund was Director of Business Development, a job later held by Nick Cirillo when Orv retired. Roger Feulner was VP of the Corporate Technology Office, which managed external technology and the India Software group, and all of which passed to Rick Borken when Roger retired. Importantly, Ben Simmons was VP of HTC, the $100 million crown jewel of the company's advanced technology, and Larry Welliver remained VP of the $100 million Solid State Electronics Center. Later I would add another function, Director of Intellectual Property that was ably held by Jeff Groat. Pat Hegstrand was my Communications Manager who proofread and improved all my writing and speeches. I mention Pat, whom everyone loved, because at one staff meeting she started to have speaking problems, which seemed odd for such a bright and competent person. Only a few months later, she died from a brain tumor. I still often think about what a wonderful person we lost.

Overall, 1993 was an amazing year. About 1,200 people reported to me, a number that would grow to 1,800 by 1999. I was an officer of the company, so my photo appeared on the back of the annual reports. My salary was among the top ten of the company. I was about to begin the most productive and busiest period of my life.

India and Poker

As mentioned, Roger Feulner had the challenging job of setting up a software operation in India. Only a couple other U.S. companies had tried such a strategy in 1993. A pair of our engineers native to India had been encouraging the idea and hoping to head up a new group there. Thousands of English-speaking software engineers and coders were coming out of Indian universities and their costs, including overhead, were as low at $20 per hour, maybe 10 times lower than U.S. engineers. Roger did an excellent job of deciding where this small operation should be located and choosing its leader. Krishna Mikkilineni, who had a PhD in both electrical and computer engineering and was working at a Honeywell Phoenix business unit, was chosen for the job, a big disappointment to the HTC people who had championed the idea. But Krishna was a wonderful choice and soon our fledgling organization was running in Bangalore, India.

My first trip to India was in 1994 to visit our new small group and encourage them. Our plan was to grow the operation to 200 people who would build software in cooperation with software design engineers in America working on company products. After a 20-hour trip through Amsterdam and New Delhi, I arrived in Bangalore. The leg from New Delhi on India Airlines was excruciating—a straight-back plane seat with little cushioning and terrible snacks. But I made it and was put up in the Taj Hotel, an old British luxury station post from the 1800s with beautiful gardens. The trip was looking up.

At that time Bangalore suburbs were exactly like the U.S. Old West, with dirt roads, unpainted wood buildings going up everywhere, and wooden sidewalks at the edge of the streets. My first ride out to the Honeywell India Software Operations (HISO) included roads where a hodgepodge of people, bikes, motorbikes, dogs, cows, buses, and cars shared the road, more or less randomly. My driver drove recklessly fast through the clutter to a small building on the outskirts of town. HISO's location was near a satellite relay tower, so they could communicate directly through the internet with Honeywell in America.

The small crew of recent hires, perhaps 30, each had a computer station. A good proportion of the employees were female, maybe 20 percent, all dressed in bright saris and typing away on their keyboards. At a small meeting, Krishna explained how their goal was to achieve the highest software (SW) quality levels in the whole company, SEI Level 5. The Software Engineering Institute (SEI) at Carnegie Mellon University developed a capability maturity level measure that became widely used in the computer industry. At Level 5 the development process for SW is repeatable, reliable, efficient, and also has continuous improvement built in. No one in Honeywell was anywhere near Level 5.

Reliable electrical current in India was rare, and it was important for everyone to know when to back up their work or to shut down their computers. There was a large sign on one wall of the computer room with three lights to indicate the current electrical status. Green meant okay, yellow meant the current was intermittent or on generators, and red meant the power was about to shut down or already had. The power failed while we were meeting. Internet connections were equally unreliable.

Krishna took me to dinner at his house that first night. A small Hindu shrine adorned a nook in their living/dining room. I tried to remember all the eating rules I'd been briefed on while at the dinner table with Krishna and his son, like not holding food in your left hand. His wife and daughter brought many servings to the men at the table. While we ate a late dinner, a few engineers chatted in an adjoining room, talking to people in the U.S. about their work. Krishna had dedicated a room in his house so his people could communicate across the 12-hour time difference.

Before leaving I took a photo of the small HISO team at their facility. They loved their jobs and clearly respected their leader. They soon achieved SEI Level 5.

Apparently we made a good choice for the new leader. Krishna is now Senior Vice President of Engineering and IT for all of Honeywell. HISO later became Honeywell Technical Solutions and now has over 9,000 employees in three countries. They even support 11 learning centers for science and mathematic teachers in India.

Some might say that the growth of these overseas jobs makes it tougher on U.S. engineers, or at least depresses salaries here. But work and progress aren't a zero-sum game. Honeywell developed an exceptional reputation among Indian engineers, largely due to Krishna and his flat organizational structure—he knew all of the employees and they knew him. Also, Honeywell was one of the first in Bangalore, right after HP and IBM. Krishna also picked excellent leaders as his organization grew. The international resource we created in India is now a major competitive advantage across Honeywell businesses that crucially depend on software. Jobs were created in both India and in America because we once took a risk in Bangalore.

My second trip to India was in 1996 and already HISO had outgrown its original facility and was building a new one in downtown Bangalore. Roger Feulner and I went to inspect the progress on the new three-story site. Building in India has major pluses and minuses. The entryway to the new building had an ornately hand-carved wooden desk for the receptionist that we were told was inexpensive. It would have cost a fortune in the U.S. On the other hand, to usher this building through all the government regulations, ensure power and internet connections, etc., required a full-time employee to bribe the appropriate officials.

Before leaving Bangalore, I was honored once more. At the airport, HISO people handed me a large box that weighed about 35 pounds. The airports there are tricky enough without an unexpected extra package, so I begged them to ship the box to my home. The gift was a stone sculpture of a young girl with a bird on her shoulder, and a three-legged table with carved elephant legs. The gift has an honored place in our home today.

I flew from Bangalore to New Delhi where Honeywell had a scheduled Asian General Managers Meeting. All of Honeywell top management seemed to be there, staying at the Taj New Delhi. The first morning of the meeting, we found five brightly clad elephants waiting outside the reception door. Apparently, the Indian cricket team was also staying at the hotel before their big game with Pakistan. The elephants were part of the pomp and celebration for the match. The Pakistani and Indian teams, hated rivals, hadn't competed for many years so someone wanted to get the home team keyed up.

The Honeywell meetings were interspersed with fun excursions set up by the company. On one, we traveled by bus to the Taj Mahal, a hair-raising experience. Each bus had two men in front, one to drive and the second to put his hand out to urge cars we had passed to slow down and let us back in the original lane before we smashed head-on into opposing traffic. We had many close calls in the thick traffic that included every manner of beast and machine. Once we had to drive through a field around a traffic jam where walkers had been killed. Our tour guide said "not to worry," since it was a normal occurrence.

We drove through villages where perhaps 10,000 people filled the streets for some celebration and visited several ancient sites before arriving at the Taj Mahal in Agra. It is a uniquely awesome monument, one of the top ten wonders of the world. Everyone has seen pictures, but actually being there was amazing.

In some ways the drive back to our hotel was equally affecting. At one stop we saw young boys diving into a pool covered with perhaps a foot of scum. Afterwards, covered in scum and stink, they scrambled up a hill to the tourists for a reward. The whole scene was horrible and very confusing. Should I give them a dollar, probably half a day's average salary in India at the time, thereby encouraging such a disgusting show? I think I did.

We passed small buildings constructed entirely of dried dung, the owner proudly sitting on his heels in front. I knew dung was used for heating there but never imagined buildings of the stuff. So many of the people sitting along the road seemed lifeless, with nothing meaningful to do. At a nice restaurant stop, I had to buy toilet paper from a man in the restroom. As the sun began to set, we passed hundreds of waiting trucks that were not allowed to enter New Delhi after 8 p.m., so we again drove off the asphalt road along the fields. For miles, sheds lined the road, each with one fluorescent light. Each small shack had a specialty, like owning a motor, doing haircuts, or having hand tools, with a small living area behind the shed.

The whole experience left me totally fascinated and yet very disturbed. The life, colors, and variety were amazing. On one special holiday during our visit, everyone on the streets was covered with bright chalk dust. Yet the poverty and pollution and waste of human life made me very sad.

The final night at the hotel, booths were set up out back with jugglers, acrobats, and food stands from around the country. We all wore turbans and sampled the food. During meetings the next day, I couldn't eat lunch—a food bug had hit me. A few others were also a little green and skipped the last lunch.

Everyone headed back to the airport later that day to take one of the three Honeywell jets in various directions. After a long delay, where cows meandered across a busy intersection near the airport, I climbed into a small Cessna Citation with my boss, Larry Moore, and a couple others. We had a short ride to Bahrain for refueling and to avoid Iran before taking off for Ireland. That's when I really got sick, sweating and vomiting in the toilet before trying to sleep. Meanwhile, Larry and two others had started a game of poker. Larry woke me and said I needed to join them. So with a terrible headache as well, I let them extract all the cash I had, about $150. Larry then kindly said that I could go back to sleep. That's how it is at the corporate pinnacle.

We had a last refueling overnight stop at a hotel near an old Irish castle before flying home. I convinced them all to check out the castle and my boss enjoyed sitting on the throne chair while we took photos. Larry Moore always said that he believed the ideal form of government was a benevolent dictatorship, as long as he could be in charge. That's the way he led his part of Honeywell, with power, intelligence, and a wry sense of humor.

Movie Star

Soon after becoming CEO, Michael Bonsignore created a movie describing his goals for the company. Then about every four months other top functional management spoke of similar plans for their areas that were distributed as DVDs across the global operations. Soon, it was my turn.

I had been boiling down my goals into just a few imperatives: (1) accelerate new product introductions, (2) greater web use, (3) increased software and product quality, and (4) incorporate wireless technology, etc. Those goals had been announced in company newsletters, but I was told there had to be a technology video. No one trusted me to pull that off alone, so right away I had to start seeing a media consultant to learn how executives should act and talk. After several sessions studying how to use expansive gestures and smile appropriately, I was ready. The plan for the movie was argued about and triple-checked as the filming day neared. Someone arranged a $100 haircut at the downtown location where all the local news anchors went. You should understand that I had not had an official haircut since my wedding. I just do regular trims with scissors by myself.

On the day of the filming I wore my best suit. Right away they told me that my favorite tie was unacceptable and showed me three to pick from. After makeup and general preening, we did half a dozen dry runs. All of this took place in a Honeywell Industrial Systems demonstration room with large display screens and equipment along all the walls. I had never seen a professional movie crew before, perhaps ten people, with cameras mounted on rail tracks across the room and special lighting everywhere. While talking about technology, I needed to walk across the room toward a large camera moving in parallel along the tracks. I pointed and held up props at just the right time. The whole film was about five minutes long. After several takes, they let me go home to wonder about the whole affair.

DVDs went out for viewing by the 60,000 Honeywell employees and I was astounded to learn that people everywhere had seen it. In Australia and Scotland, for example, engineers had memorized my goals, using the exact words. That's when the power and influence of my new position hit me. When I spoke, people assumed my words were approved company policy statements. That might have made me cautious, but I started to see the potential. I didn't really need to ask my boss or his boss for permission generally. As long as I had the position of VP of Technology, in front of any audience I could just lay out my vision of the future and people would actually listen. Cool.

Ups and Downs—Three Wars

Even though Honeywell had sold off its bombs and torpedo businesses, it was still fairly dependent on the government for revenue. So after the Berlin wall came down, there were a couple of years with a "peace dividend" for our country but reduced income for Honeywell. Those were the years when Chris Steffen drove layoffs, selloffs, stock buybacks, and legal tricks to keep Honeywell's stock price up. The stock market, by then, expected about 15 percent profit growth per year from big companies.

Then the first Gulf War began with Iraq invading Kuwait. The military and space businesses picked up speed, but mostly for short-term fixes. Research in Honeywell labs dropped as funds were diverted to the ongoing war. Also, winning the 777 aircraft development contract was great, but it required about $120 million of investment from Honeywell. The first years in the 1990s were financially very tight.

A basic rule of business, at Honeywell at least, was that "there is always a crisis." Several of the 20-plus businesses were always in trouble, and the successful businesses had to make up the slack. So around 1994, every healthy division had been taxed (given a higher profit goal for the year). That's when I made a bold, risky move—I actually asked for more money.

HTC was receiving about $30 million of internal company funds at that time, and I went into Larry Moore's office alone and asked that it be increased to $33 million. Here, my high empathy quotient was indispensable. I knew him very well and understood all the things he cared about, so I proposed several perfect new programs. The worst he could do was turn me down. One was for situational awareness technology for military and commercial pilots (including some work on helmet-mounted displays). Another was to start a software initiative for the entire company, to raise the quality standards to SEI Level 5.

By focusing on new programs he liked, he actually agreed to the new funds for HTC. Soon additional taxes were levied against all the business units to get us the new money—and they were very angry.

Perhaps this was the final challenge for an introvert who wanted to fly—to learn how to talk directly to people who were furious with me. I started a systematic tour of the business line managers to apologize, explain why the new programs were so great, and to ask how we could help them more in the future. One general manager said he understood, but that to him all R&D was a tax, and he hated all taxes. Overall, I think the tour was a success because I looked everyone in the eye. Most important, I listened and they knew I understood their problems.

The Bosnian War came to a head in 1995 and required additional military innovation, especially for smart munitions. The Boeing 777 was finally in production, so profits had begun to flow to Honeywell again.

A third war began in 1999, the Kosovo War. The Serbs and Croats were waging a war of genocide against the Muslims in Kosovo, and the U.S. began an air campaign in conjunction with NATO troops. During the first Gulf War, vast numbers of dumb, 1,000-pound bombs had been dropped on the Red Guard with accuracy of a few hundred feet. It was incredibly wasteful and caused tremendous collateral damage. But 1999 was finally the age of smart bombs.

Remember that my science area developed very small ring laser gyros, starting in 1983. By 1995 they were finally in production in a program called Joint Direct Attack Munition (JDAM), which used both our small gyros and the Global Positioning System (GPS) for targeting. The gyro was needed because it is possible to jam the GPS near valuable targets. They were inserted in the dumb, old bombs, with tail fins added for guidance, and suddenly the accuracy had improved to four or five feet. An entire new building was constructed by Honeywell to produce the gyros.

I mention all this due to an interesting story at the end of the Kosovo War bombing. One night the Chinese embassy was hit with five JDAM bombs, which was odd because of their great accuracy. The official word was that a building across the street was the actual target but the wrong coordinates had been used. Most of my friends believe that the targeting was on purpose. China had been supplying arms to the Serbs during the war, and a relative of the Chinese premier was working at the embassy. The bomb hit late at night and three Chinese reporters were killed. The U.S. later compensated the families and paid for the damage to the building. But the CIA had sent a message—stop supporting the Serbs.

Despite all the small wars, the late 1990s were great years for Honeywell and business in general. People were finally using the technology of the internet, wireless, and fast computers in innumerable ways to improve lives and the dot-com bubble was still growing.

Morality and Military Technology

It should be clear that during the '80s and '90s I had both a direct and indirect role in creating new military capabilities. The JDAM missile just mentioned is a clear example that led to many casualties. Their precision, though, led to far less collateral damage than would have previously occurred. Stopping a war that led to the forced expulsion of 800,000 Kosovo Muslims and the killing of about 10,000 was a valid intervention. There are people bent on evil who must be faced occasionally.

Our lab also had a major role in the development of the stealth capability in the 1980s. Stealth aircraft require many advanced technologies: radar-absorbing materials, flat/angled surfaces, and new flight control techniques. Jets with flat surfaces don't fly well. S&RC had many scientists behind closed doors developing the fly-by-wire equations for both the stealth fighter and bomber. Stealth, *Star Wars* technology, and the internet were key reasons that Gorbechev finally gave up, ending the Cold War and leading to the dissolution of the Soviet Union and the removal of the Berlin Wall. This was very good.

I have very mixed feelings about science and the military. My beginnings at Honeywell, working on solar energy collectors, had drifted over the years to less benign uses. Still, generally, I'm proud of the work we did. Many of my thoughts and concerns about the duality of science, for good or for evil, continue today. Will humans learn to balance and control their consequences? I firmly believe that over time they will.

Math, Walking Fish, and Submarines

Our daughter Amy's high school years were successful and interesting. An excellent student, she graduated in the top ten of her 500-person class. Mounds View High School ranks as one of the top public schools in Minnesota, both academically and in sports. She was one of eight National Merit Scholarship Finalists in her class.

In fact, she did well in many areas, with the exception of athletics. One summer she finished her fifth year of Swedish camp at Concordia Language Villages near Detroit Lakes, Minnesota, and received high school credit. Photography became a passion. She developed her own photos and even ventured out in an ice storm to get close ups. She traveled to Paris and London with the school band, playing flute. On the side, she taught Sunday school for three years. By her senior year she was involved in national politics and attended a Wellstone training camp.

Throughout these years, mathematics stayed as a key focus. A very special teacher, Dan Butler, who had coached the math team at Amy's middle school, began teaching at Mounds View and quickly took over the Math Team. Mr. Butler mentored and encouraged all the students on the team, but four were special—Amy, Nathan Doble, Jenwa Hsung, and Michael Korn. The team consistently scored among the top three in Minnesota through Amy's high school years. Michael Korn was the best individual in the state two years in a row. Both Nathan and Jenwa ended up at MIT and Michael studied later at Princeton.

Despite my rule that Amy not date anyone until she had finished graduate school, she completely ignored me—dating both Nathan and Michael. On New Year's Eve our family always stayed home for a night of competitive games that ended with everyone shouting "Happy New Year" into the suburban night at midnight. Nathan and Michael joined us on successive New Year celebrations. I especially tried to beat Michael Korn at every game, knowing it would be a challenge. All I remember is that he was very good at computer games. Michael eventually competed in national and international math competitions, ending up 6th in the world representing the U.S. in India. He now works for the National Security Administration (NSA), breaking or writing codes probably.

In the summer between her sophomore and junior years, 1995, Amy and I made our first college visits. She checked out Harvey Mudd and also stayed overnight at Dabney House at Caltech. We would visit more colleges the following summer.

She also packed her bags that summer and we drove her to Rose-Hulman Institute of Technology in Terra Haute, Indiana—to Math Camp. Top students from all over the U.S. stayed there for six weeks to get a taste of advanced math and to see how other smart kids act. Some of them were a bit odd and a few got kicked out early for staging a dorm raid. But Amy also met students who had been home-schooled, attended elite private schools, or public schools with focused science and math curricula. She was startled by how clever and intense they all were. Her experience echoed mine when I attended the summer chemistry program at Augsburg College. It's good to size up the competition early and adapt.

The next summer we visited more colleges, including Rice in Texas and, on an East Coast tour we visited Carnegie Mellon, RPI, Cornell, Brown, and MIT. At Brown, the hills near the school had been set up for downhill X-Game competitions with bales of hay everywhere—impressive. Amy applied to all eight universities and was accepted by them all!

She ultimately chose Caltech. She was attracted to it for many of the same reasons I had been many years earlier. Caltech was the most difficult school to get into, its students had the highest scores on tests. They would pose the greatest challenges and would be fun to know. Caltech has a beautiful campus. A small college, it's very supportive of its students and has a surprisingly broad curriculum. I think the fact that I loved it didn't hurt her decision.

As early as 1995, I knew Amy was leaning toward Caltech. I also knew that just getting good grades would not be enough to get in.

She needed an impressive scientific project of some kind. I also knew that muskrats were burrowing into the shore of the pond by our house each year, slowly eroding the bank. Occasionally I would see them swimming in the early morning and it really bugged me. Discussing this at the dinner table, Amy and I decided to build a submarine. It would help her get some engineering experience and I could chase the muskrats away from our shore. Miriam didn't yet understand the consequences of this decision, or she might have objected.

The sub became a very cool summer project. To get her friends involved, we produced a slick brochure showing all the fun they would have and the experiments they could do. I envisioned a two- or three-foot-long submarine with remote control. Amy took it from there.

She got the word out to her UMTYMP and Mounds View friends and some new people she had met in a science club called the Walking Fish. The Walking Fish were good students in the Golden Valley/Hopkins area near Minneapolis who had been encouraged by some parents, the Borcherts, to do fun activities together. The Walking Fish included Yishan Wong (recently the CEO of Reddit), E. V. Möebius (related to the inventor of the Möebius strip), and a floppy-haired (like the early Beatles) boy named Dan who played an increasing role in Amy's life. Altogether, about 10 teenagers regularly started showing up at our house to work on the sub while I took on a consulting role.

The submarine began to grow, both in physical size and complexity. Soon, my small sub idea had morphed into an 800-pound, nine-foot-long monster. Constructed with wood framing, it included two electric trawling motors, a large car battery, a ballast tank with compressed air to blow out water, a controllable back fin, and electronics to control both the fin and solenoids for the ballast.

As the sub grew in size, more weight was needed to get it to actually sink. I purchased many buckets of lead wheel weights from a local salvage company, each with little steel clips. When heated, the lead would melt and the steel clips would float to the surface. One night around midnight, the students were still working and melting the lead weights over our Weber grill in the middle of our backyard. They were playing loud rock music and scooping the molten lead and dropping it into the bottom of the wooden sub. One girl screamed, "Stop splashing that lead on my leg!" The group was dancing around the flaming pot, fanning it with cardboard and singing along with the music. It was an image I will never forget; it seemed like a high-tech pagan ritual. I can't imagine what the neighbors thought.

Later, a professional came over to fiberglass the outside of the sub while it hung from a tree like a gigantic fish. It also looked a bit like a Transylvanian coffin. A window in the front held a camcorder and some sensors. An access hatch sat on the top. The electronics inside were a rat's nest of wires, despite my suggestion that they be better organized. I argued with some of the kids about how to control all the actuators while underwater. Some thought that our model airplane controller RF would penetrate the water, but I knew it wouldn't due to salts and dissolved minerals. So eventually a wire line was added to communicate to a surface pod and then to shore. A trailer had to be purchased so the whole sub could be moved to a lake. On the final days, Yishan Wong painted aliens on the side of our yellow submarine.

I learned a lot about kids that summer. First, they are very messy—our basement carpet was ruined with clumps of epoxy stuck in the nap. Statistically, the average teenage boy eats 6.3 tacos for lunch when given the opportunity. And I noticed Amy was hanging out with Dan, the floppy-haired boy, a lot.

School had started again by the time the sub was ready for its first dive in September. Amy and others contacted all the local media. Our small pond had poor visibility, only about two feet. So we towed the sub to Turtle Lake and set up a TV on the shore so people could see clear camcorder images from underwater in real time. A large crowd gathered, including TV and local newspaper reporters. Amy was interviewed and then christened the sub by pouring soda over the outside before it was lowered into the water. The TV images showed seaweed whipping by as it moved under the lake surface with teens swimming nearby trying to follow along. I was nervous about going too deep and having the whole thing implode, leaving 800 pounds of toxic lead at the bottom of the lake. So we didn't go too far down, but the submarine actually worked. The tail fin broke, but the ballast tanks functioned.

The newspaper clippings were included in Amy's college applications, and maybe in some of the other kids' as well. I was left with a nearly immovable object blocking the path to my lawnmower and a lot of great memories. Chasing muskrats in our murky pond never happened, but Amy had found a new boyfriend, Dan Fisher.

Invitations

When one reaches the top levels of a major corporation, invitations start appearing from everywhere. I already mentioned two University of Minnesota organizations, the Institute of Technology (IT) and Center for Development of Technological Leadership (CDTL) advisory boards, which I contributed to. I am still, in 2018, on the CDTL board (now called the Technical Leadership Institute [TLI]). For a while I played a similar role on an MIT advisory board. During the 1990s, the Minnesota High Tech Council also invited me to join.

One very impactful study group, called Future Grand National Challenges, met at the Center for Strategic and International Studies in Washington D.C. It drew technical leaders from across the U.S. and was headed by Harold Brown, former President of Caltech and Director of Research for the Defense Department. Politics and persuasion at that level is very interesting. I'm afraid my introversion kept me from having a major impact on the study results. A similar study by the Rand Corporation that I participated in was widely published and used for planning throughout government for several years.

Two other invitations were especially interesting. In one, an organization called the Minnesota Technology Corridor Corporation (MTCC) asked me to help in their goal to develop the area along Washington Avenue and the Mississippi River in downtown Minneapolis. A supercomputing center, a technology incubator building, and a seven-story tower had been built in that area since its founding in the 1980s. The general idea was to jump start new technical companies in the region near the university and downtown Minneapolis and to clean up the Washington Avenue area (previously a seedy area for drunks). But many companies had declined to build new buildings, including Honeywell. The Board of Directors for MTCC included the President of the University of Minnesota (Nils Hasselmo), the former Minneapolis Mayor (Art Naftalin), the former CEO of ADC Telecommunications (Chuck Denny), the former President who helped found MTS Systems (Herb Johnson) and many others. Herb became sort of a mentor to me during those years and ultimately asked me to become Chairman of the Board for MTCC. But by the mid-1990s it was becoming increasingly clear that the location for technology companies wasn't too important because of the rapid growth of internet technology. So my role became mostly to oversee the development of park space along the Mississippi and to slowly shut down the whole nonprofit company. Since then, more general-purpose offices and apartments have sprung up in the area. But at least I had the chance to be the Chairman of the Board for a while.

Another invitation, actually a directive from Honeywell management, was to become involved in what is now known as First, but which was then

called U.S. First, a national competition among high schools in robot build-
ing and games. Minnesota had a few teams competing at the time, and
Honeywell was assisting the North High School team. U.S. First wanted
Michael Bonsignore to join their corporate leaders board, but I was stuck
with the task. U.S. First was led by the infamous independent entrepreneur,
Dean Kamen. Kamen had invented the infusion pump for medical applications, home dialysis technology, an advanced prosthetic arm, and later the
Segway human transporter. Each year, after the robot competition in New
Hampshire, he would award a traveling trophy to a corporate supporter—a
large transparent plastic clock, several feet tall, that he had built. The beautiful clock was destined for Honeywell one year, so I was asked to give a
speech to all the students assembled and take possession of the clock.

It turned out to be a wonderful weekend. Each team was given a box
full of parts with which to build the robots, and the competition was carefully defined before they started building. As an official judge for the robot
battles, I got to see the excited builders use their robots to throw balls into
a hoop and try to prevent their competitors from doing the same. All the
while, school cheerleaders were creating a frenzy. Kamen had designed the
competition to resemble a high school basketball game in a large gymna-
sium. It was a lot of fun.

Since then, the popularity has spread worldwide and hundreds of schools
compete for a chance to be on national TV for the finals.

On the final night of the competition, all the winning teams received their
trophies and I stepped up to get our clock. On the way, Kamen whispered
in my ear, "Keep it short. You don't have a long speech, I hope." Fortunately,
my speech was short, but then he rambled on for 30 minutes. Kamen is a
very self-centered man, which is one reason I later turned down a chance to
fly on a helicopter to his home and laboratory on an Atlantic island he owns.

The beautiful clock, however, stood proudly in Honeywell's corporate
lobby for one year.

Company "Meetings"

Two- or three-day meetings with other general managers occurred annually,
as mentioned earlier, and the level of extravagance depended on the financial

results the previous year. Under Renier, they were mostly in Scottsdale at a new resort each year. Bonsignore moved them around more. Miriam learned that despite her teaching responsibilities such opportunities should not be missed, although I think she took along papers to grade each time.

We were in Palm Springs one year and I was expected to play golf on the Indian Wells championship course. Honeywell took over the course one afternoon, with 30 or 40 golf carts lined up. Each cart basket was filled with new balls and special gifts for the participants. We also enjoyed a hot air balloon ride, with gorgeous views of the city, desert, and mountains. The ride ended with an exciting crash landing on a vacant city lot among sage and cacti. The basket tipped over on landing, but no one was hurt.

A meeting in Key Largo, Florida, occurred after a particularly profitable year, in February 1996. The entire management team, with spouses, took a cruise on an immense yacht where we danced to a band on the top deck. Before that meeting ended, Honeywell spent over $100,000 on fireworks the final night. Everyone gaped as the seemingly endless display exploded over the Atlantic beach, before the finale where sparkles attached to scaffolding spelled out HONEYWELL about 50 yards offshore.

One of the GM meetings took place at Disney World, but the really decadent outings were with the Board of Directors (BOD). I was invited to those every second year, when long-term strategy was the main subject. One at the Boulders Resort in Arizona was memorable, with the first GPS-sensing golf carts. But the BOD meeting Miriam and I will never forget occurred at Sea Island, Georgia, in 1994.

I picked Miriam up after she'd experienced a particularly difficult afternoon teaching at Como Park High School. We left my car inside the Honeywell hangar and took the company's Cessna Citation with Larry Moore and his wife. At the Georgia landing site, two limousines were ready to take us to the resort. I guess they didn't want to mix two levels of management in only one limo. At The Cloister island resort, we were simply shown to our personal multi-room cottage without the need to touch a bag or check-in at the front desk. The suite we stayed in would cost about $2,400 per night today. It was stunning. Typical guests at The Cloister include presidents and other heads of state.

Scheduled for my first long talk the following morning with the BOD, a dry run with my slides seemed a good idea, so I got permission to enter the presentation room early. Bonsignore's secretary had demanded that all the slides be 35mm film wedged between two thin glass wafers. This was a technique used then to allow projection with powerful bulbs. That night I discovered that moisture had condensed inside all my slides, a result of coming from the cold airplane to the hot humid Georgia air. When projected, the water evaporated, causing air pockets shaped like worms to crawl over the projected image. The effect was very annoying and distracting. I warned everyone about the problem at the dinner that night but no one seemed to care. In the morning, I placed my slides into the warm projector early to drive off all the moisture, and it worked. Since I was there early, I had a chance to watch Honeywell's security chief scan all the furniture and walls with special equipment, looking for electronic bugs. That's how important those meetings were. After introductions, my talk was scheduled first, and it went off well with no slide problems. The room was filled with powerful people, including the heads of Northwest Airlines, SuperValu, and Xerox. I later became quite friendly with the SuperValu guy, but the main thing was that I survived that first talk. The rest of the day, they all watched the other speakers as little worms crawled across slide after slide. I resisted the temptation to say, "I told you so."

While the Honeywell managers and directors held meetings, the spouses went on tours of the island and to special events. Miriam was disturbed to see that children from former plantation slave families had been pulled out of school to sing and dance as their grandparents once had.

She also got to listen to the gossip and comments of the other wives. One said that when her husband didn't pay her enough attention, she'd go out into the Arizona desert near their home and gather up burrs. Hiding them in her husband's underwear got her message across.

The final dinner proved to me that many of the executives were like aliens of a different species. Around the dinner table they were astounded to learn that we didn't own at least three houses. Some of them had five. An argument started between the largest real estate company owner in California and the CEO of the largest bank in Canada. They jointly decided that we needed to buy a new home in Arizona but quibbled over the mortgage rate the Canadian could charge.

Meanwhile, wives at our table talked about recent trips. One had just returned from a helicopter skiing outing in the Andes and another described her weekend shopping trip to Paris.

I worried that Miriam would start to argue with Bonsignore when the state of the school system came up. She held her own and didn't get me fired—yay! Generally, the contrast between teaching inner-city teenagers and eating with the corporate privileged was a bit of culture shock.

Bosses and the Fifth Floor

Being promoted to lead the company's technology entitled me to a third office at corporate headquarters just off the I35W freeway south of downtown Minneapolis. In the Camden R&D building I had a third floor corner office where I could work hard, or daydream while watching the large train yard nearby. I'd picked an oval conference table and nice couch for that room, and Carol Warne protected me from any unwanted guests. When I was at the Plymouth facility I had a nice workroom that doubled as conference space when we weren't there. But my new corporate office was amazing. Carol sat in a large entrance chamber with a half dozen file cabinets and chairs on which people could wait. My personal office was at least 16 by 20 feet, with furniture and paintings I selected from several catalogs. A wall of windows overlooked the company plaza and the freeway.

Honeywell's corporate 5th floor was infamous throughout the company. Stepping off the elevator, you were confronted by a receptionist. CEO Jim Renier had installed a wall of bulletproof glass during the period of layoffs and the receptionist needed to push a hidden button for the large glass wall to slide open. The fifth floor had about a dozen giant offices like mine, or larger, and it was normally very quiet there. The executives were constantly traveling and 99 percent of the time one could roll a bowling ball down the length of the long hallway and no one would even notice. Because it was such a sterile atmosphere, Carol Warne and I spent as little time there as possible.

A common trait among all the people on the fifth floor, except me perhaps, is that they could be simultaneously friendly and scary. You would love to have any of them as a neighbor but would be terrified to cross them or let them down. Over my career at Honeywell I had nine distinctly different immediate bosses, including:

- Jaan Jurrison—warm, encouraging, father-like
- Henry Mar—friend, mentor, thoughtful
- Mel Geokezas—intense, hopeful, insecure
- John Dehne—very ambitious, smart, quick
- Carl Vignalli—calm, lenient
- Clint Larson—bureaucratic, supportive
- Len Weisberg—intense, brilliant, critical
- Larry Moore—strong, wise, loyal
- Giannantonio Ferrari—hard to understand, prideful

To advance up the corporate ladder I had to constantly adjust to different personalities. In some ways, I would imitate their behavior a little and try to be close and unique to them. Being an introvert really helped in understanding each one, analyzing their behavior, and being empathetic to his needs. In an early discussion with Len Weisberg, he described how each of the people who had run S&RC had dealt with him. He said Roger Heinisch kept giving him projects to work on and ignoring him, and how John Dehne had constantly gotten into fights and argued with him. He said he wondered how I would be. Over time, I tried to be his friend and he grew to respect me. Len had a reputation as a tyrant, and I had watched him humiliate one engineer who came to speak at a Ring Laser Gyro (RLG) mirror task force meeting. Before the engineer could speak, Len said, "I hope you are not here to ask for more money." After two or three sentences, the engineer said something that implied he needed funds and Len kicked him out—in front of about 40 people.

A higher-level boss with a tough reputation was Ward Wheaton, head of the Aerospace group. He had nine children and directed each one to write a five-year strategic plan for their lives each year. One of his sons, David Wheaton, became a well-known professional tennis player. For my first presentation to him, he treated me well, to my great relief. I tried to exude honesty, and I think he considered me a "troop," not a lieutenant to be reamed.

Being willing to talk about bad news honestly is very important in management. That approach also helped win over Len Weisberg. He held very strong opinions about work and people, but once shown new facts, he could instantly change his mind.

My own management style was one of putting almost everything on the table. I encouraged people to lead, brainstorm, and try out new methods. Everyone was asked to spend at least 10 percent of their time learning or improving their job, not just doing it. I often used the example of a soldier at the Alamo facing hundreds of charging enemies. There was no time to talk to the machine gun salesman at that moment, even though it could have saved the day. They needed to do that earlier.

Here's one final story about my favorite boss. At the corporate offices we parked our cars in the heated underground garage. The room was filled with beautiful, clean, expensive cars. An attendant would wash each car every third day and, as mentioned, we all received a new car every third year. I owned mostly Porsches and BMWs during the 1990s. Every time I got a new car, my boss, Larry Moore, demanded that I let him have it for a while. He would trade one of his, like his wife's Lexus sports car, and take mine for a weekend. He loved motors and had his own airplane, motorcycles, boats, and fleet of 10 cars. Larry retired in June 1997 and I was sorry to see him leave. He had an internet blog where he talked about lovingly polishing his Mercedes. Sadly, he died not too many years later.

If Only . . . Osheroff's Nobel

While driving alone in 1996 in the western suburbs of Minneapolis toward the Spring Hill Conference Center where I was scheduled to give a talk, I heard someone on the radio say something about Cornell. Listening more closely, the announcer said the Nobel Prize in Physics was being awarded to David Lee, Robert Richardson, and Douglas Osheroff for their discovery of superfluidity in helium-3. I distinctly remember saying aloud, "Oh no," to myself in the car. My college roommate, Doug, was getting a Nobel Prize and sharing about $900,000.

Suspecting for some time that he might win the most prestigious award in science, it wasn't a total surprise, but it did lead to comparisons and reminders of our time together. If only my refrigerator had been better and colder. If only I had been looking somewhere no one had searched before, instead of trying to solve an existing problem. Living off campus with Osheroff, I always thought I was a bit more clever, but he clearly was a harder worker. If only I had studied a little harder at Caltech.

I was having fun at Honeywell and making a ton of money, but my roommate would have his name engraved forever with some of the most important people of our lifetime. That's how things go sometimes.

Australia, Singapore, and Bora Bora

My first trip to Australia turned out to be a frequent-flyer bonanza. The engineering and management people there had invited me to come see their new product lines. People from New Zealand had also flown in for the meetings. As I mentioned, I was astounded to find that they had virtually memorized my technical goals for the company and wanted to show off how they were using them.

But before traveling to Australia, the airline called and asked if I'd be willing to leave six hours later, and I agreed. Since I was already traveling business class during the off-season, my frequent flyer miles had been doubled, but then they proposed to double the miles again, to about 40,000, and also offered a free trip for one and a half people from Los Angeles to Tahiti and back. That seemed like pretty good compensation for a six-hour delay and enabled a later trip for Miriam and I to Bora Bora.

What I remember best about that first trip was the wonderful, friendly treatment by the people. They seemed like close friends after only a few days. I was invited to a BBQ (barbie) at the General Manager's house overlooking one of Sydney's beautiful bays. With prawns, chops, and lots of beer, it was a fun evening. Off work, I had time to wander through downtown Sydney and see the parks and waterfront.

My second trip to Australia was for an Asian General Managers Meeting. Arriving a little early, I took a ferry over the harbor to Sydney's Taronga Zoo to meet one of my former Sunday school students, Amanda Nelson. She was living with temporary guardians for the summer for an overseas adventure. The zoo is on a steep hillside overlooking the famous Opera House, bridge, and cityscape across the bay. Giraffes and other animals have one of the best views in the world from their enclosures. Amanda, her guardian, and I had

a great time. Amanda liked her Asian trip so much that she is now married
and living in Shanghai.

Half of Honeywell's top management came for the Australian meeting,
most on the large corporate jet. The highlight for me was a jaunt out to the
then future site of the 2000 Olympics. Honeywell Australia had responsibility for security at the Olympics, so we saw the inside of the control
center with hundreds of TV security monitors looking down on every inch
of the site, about two years before the actual games. Rail transport from
Sydney was still being constructed and we had a first view of the Olympic
stadium. After the meetings, most of the executives flew off for a similar
meeting in South America while I headed off to Singapore. It's easy to see
why Honeywell's 5th floor was often deserted.

The stop in Singapore was for an interview by a government commercial department. They were trying to establish creative research groups in
their somewhat robotic society. It's hard to nurture creative people in a pre-
programed, paternalistic culture. Maybe making it legal for people to chew
gum in public would help. I told them how we did R&D, but I'm not sure
they understood. Later, they would fund a small startup group of researchers for Honeywell's Technology Center. This group never grew as fast as
our software team in Bangalore, however. But Singapore is a beautiful and
amazingly clean city. I especially liked a botanical garden specializing in or-
chids on a high hill overlooking the city.

The big payoff from all these trips was an almost free flight to Tahiti and
then Bora Bora for Miriam and I to celebrate our 25th anniversary. Since
the airfare was mostly covered, we decided to splurge at a fancy beach resort
for about $600 per night. Our personal wooden bungalows had stilt sup-
ports over the water with a glass bottom floor area so we could watch colorful fish swimming at night with underwater lights. Bora Bora was, at that
time, essentially a communications-free zone. Our hotel had only one phone
for guests, for emergency use only. Virtually every couple there was on their
honeymoon or big-number anniversary, and they came from all over the
world.

The turquoise waters inside a circular atoll are gorgeous, while lush palms
cover the slopes of volcanic Mount Otemanu. Romantic ambiance wafted
over the warm volcanic island.

We had our own kayak for exploring the nearby beach area and each day we took an excursion: manta ray feeding, Jeep ride up the mountain, town visit, or a catamaran sail into the sunset with classical music and drinks for all the couples. We limited ourselves to one two-hour excursion, with soul-restoring free time the rest of the day. It's hard to describe the sheer splendor of this island and its restorative powers. It was the most beautiful vacation of our lives.

Futurist

I often referred to myself as the company's "futurist." There were a hundred forward-looking views available to me in our research lab, from visits to other companies (like those in Silicon Valley), on national boards, from the University of Minnesota boards, or from my readings.

"Prediction is difficult, especially about the future," was a favorite quote from Len Weisberg who stole it from Niels Bohr. But I believe scientists who know what technologies are becoming possible *can* understand the broad trends.

Perhaps my most valuable glimpses of the future came during trips to Europe for Honeywell's futurist competition, a challenge to all European college students to write a paper describing the world 15 years in the future. About 20 countries held national competitions to pick their top two candidates for the finals, with a year of free graduate school in the U.S. as the prize. The weeklong finals, involving verbal defense of their predictions, were broadcast in newsflashes on CNN in Europe. I became a finals judge for the biannual competitions.

The 1997 finals were held in Munich in a hotel only one block from the legendary Hofbräuhaus, where Miriam had gotten tipsy drinking beer with the U.S. bobsled team many years earlier. One candidate in particular alarmed me with his brilliance. He was a law student from Germany who on the side was a computer expert in contact with some of the greatest scientists in Europe. His paper was about product lines of two intelligent computer software companies that decided to merge when their programs began to overlap. In his story, the government tried to prevent the anti-trust merger, but the laws were too old and slow, and it became impossible to disentangle the valuable monopolistic programs. In the student's verbal defense, I can remember his first words, "I am evolution." He went on to describe how intelligent programs evolve faster than humans and will eventually displace our obsolete species. The whole talk was haunting and terrifying.

The other winner that year, a woman from Sweden, showed a talking computer teaching a child with speech impairments how to improve using a device in the child's mouth and cartoon computer images coaxing the child along.

On the final night, all the students invited the judges to a party at the Hofbräuhaus where we got to know them better. Two students eventually worked at the Honeywell Technology Center. We remain in contact today. It was great to meet the other judges as well. One was the technology editor for CNN and another had helped found both the *Economist* and *Wired* magazines.

Two years later Miriam joined me in Lisbon for the finals. She toured the city with other spouses, and we had one free day together to visit the famous waterfront and the beautiful town of Sintra.

Even ex-Soviet Union countries participated in the competitions and they gave me a wonderful view of young people throughout Europe.

Big Transitions for the Girls

In 1997 Kelsey began Mounds View High School and Amy went off to college. Kelsey, too, soon started ignoring my advice about not dating until after graduate school. She focused on a boy named Steve Berg, who was ridiculously romantic and giving. He sometimes came over to make breakfast for her. One time Steve and his father made snowmen that held up signs in our yard inviting her to the winter Sno Daze dance. It was surprising because there was little snow in our yard. They had hauled it all in on a pickup truck. Another time, Steve took Kelsey up in a small plane and circled our neighborhood so she could see a large sign on our roof inviting her to the prom that year.

Kelsey had a great group of friends in high school who were all "good kids." Excellent parenting can go only so far in rearing a wise and happy child. You also need to luck out with the character of the friends she chooses at school. Those friends often have a greater influence on how your children behave and mature. Both Amanda Melquist and Laurel Weiske were Kelsey's very close friends, but another half-dozen teens were often at our house before some event, and we have probably a hundred photos of these kids hanging out together.

Kelsey was an excellent student during high school, graduating in the top 15 of a large class and ultimately winning a college scholarship from the St. Paul Teachers organization. She played trumpet and was in the National Honor Society. Before high school, she had sung in four musicals, including memorable roles as Woodstock in *A Charlie Brown Story* and as the scarecrow in *The Wizard of Oz*. She taught preschool kids in a summer parks and rec program, served as a nanny and junior counselor at Girl Scout camps, and volunteered at St. John's Hospital. She did lots of Girl Scout activities and got a Silver Award (second highest possible).

But Kelsey mostly threw herself into sports: cross-country running, skiing, and track. All those years playing soccer midfield had enabled her to go seemingly forever. An early cross-country meet in Alexandria, Minnesota, shocked us all. Running with all the junior varsity kids from two dozen schools, a couple hundred girls ran off into the woods on a golf course where beautiful fall leaves were at their brightest. About 15 minutes later, Kelsey came back in first place with a huge lead. We were all totally surprised at this win. She continued to run well through high school but never again achieved such a lofty position.

As her high school years passed, Kelsey focused more on the teams, becoming captain of both cross-country running and ski teams. She excelled at organizing, supporting others, and working with the coaches. Soon she had all the kids working on little projects, like learning how to knit on the long bus rides to their meets. She even taught the male coaches how to knit and sew and generally built camaraderie and team spirit. Even today, people love to be with Kelsey and to work with her.

During her junior and senior years, Kelsey and I visited potential colleges, Pomona and Cal Lutheran in the L.A. area, Trinity College in San Antonio, and Valparaiso (Valpo) in Indiana. She, too, was accepted at all of them, except Pomona (which she claims is because a friend from her high school was a championship diver and was accepted over her). The overnight visit to Valpo stood out for her because, as she said, "Those kids are super friendly." Kelsey would start at Valpo in September 2001 in the Christ College program, limited to the top-potential students.

I had flown with Amy for her start at Caltech in 1997. She was given a temporary room in Ricketts House (as I had been 30 years earlier). Next door was a girl named Cheryl Forest who became Amy's roommate and a lifelong friend. As mentioned, Caltech allows all manner of construction and "personalization" of the student houses. So, compared to access holes in the ceilings that students had made leading to the next floor up, painting the walls is small stuff, as you can imagine. Elaborate and sometimes brilliant drawings and writings covered the hallways and stairwells. Lewd and rude depictions or sayings are allowed, as long as they are very clever. When I was helping Amy move in, Cheryl Forest's mother came through and was obviously shocked by what she saw. She even threatened to pull Cheryl out of the school. I talked to the mom about my experience at Caltech, and how the good far outweighed the sometimes-raw environment. There were peers for Cheryl and Amy who would challenge them in every possible way: socially, academically, athletically, and morally. They would grow into strong, self-assured adults very quickly at Caltech and have more meaningful lives as a result. I didn't say all that to Cheryl's mom, but enough to let her daughter stay at the school.

After getting Amy settled in, she walked with me to our rental car parked on California Boulevard. A big hug later, with tears in her eyes, I pulled away, watching her in the rearview mirror. I was teary also while driving back to the airport.

Amy adapted to Caltech quickly, attending the freshmen camp which had been moved to Catalina Island, and later rotating to Blacker as her permanent house. Cheryl became her roommate and they soon sponge painted the walls. Amy had an upper bunk/loft for sleeping.

It's interesting how Amy dealt with the pressures there. Unlike me, she maintained her religious connections, joining the Caltech Christian Fellowship. She avoided the social pressures of the still predominantly male school by claiming to have a serious boyfriend back in Minnesota and putting Dan's picture on her desk. By December of her sophomore year, she made good on that claim by getting engaged to Dan Fisher. Throughout her Caltech years, Dan visited her in California and she came home to Minnesota as often as possible. We bought her a new green Honda Civic her sophomore year and Dan flew to California that spring to help her drive back to Shoreview. She worked at Honeywell two summers, once at the Honeywell Technology Center and once with the Commercial Aviation Division.

Probably her biggest shock was in her first year math class when she got a D. This was after starting math classes at the University of Minnesota as an eighth grader. She probably considered math her best subject. But classes at Caltech are not normal. All the professors assume the students know all the material, sometimes before even taking the class, so they have to do problems and exams that go beyond. They examine uses that are new, combine ideas in new ways, and expect the students to generally extrapolate their learning. Also, Amy's math professor was very theoretical and possibly wanted to show the students that he was smarter. In any case, like me, Amy switched out of math. She became an Engineering and Applied Science major. One summer she stayed on campus for the Summer Undergraduate Research Fellowship (SURF) program, working with Professor Yaser Abu-Mostafa. They looked at how the stock market moves beyond reasonable changes due to human emotions. They did this with mathematical system models. I've always claimed that Amy could have become very rich as a Caltech-educated systems engineer with financial expertise. I'm sure Wall Street is looking for such people.

Amy's athletic experience at Caltech was bizarre. One year she joined the official Caltech men's golf team. Remember, she is not a man, but that didn't seem to matter much there. She got to practice and play golf at some ridiculously expensive courses in the Los Angeles area, although she was never among the top six players on the team. Her scores didn't count, but she got some excellent instruction.

Similarly she got one-on-one instruction in diving. She had signed up for a diving class, but everyone else apparently dropped out. At Caltech, expense is not a big consideration, so Amy got personalized instruction from an Olympic-level diving coach.

We came out to visit Amy a few times during her time at Caltech. In 1999, Miriam and I first attended a weeklong Sweatt Award celebration in San Diego at the Hotel del Coronado. We visited the Scripps Aquarium, the San Diego Zoo, Sea World, the Mexican Old Town area, and the Safari Park. One very cool part was getting a ride on an America's Cup yacht. Serving as a host for the Sweatt Award winners was one of the nicest parts of my job.

After the Sweatt celebration, Miriam and I drove to Pasadena to visit Amy. Kelsey and a friend flew in to join us there and to check out the L.A. colleges, beaches, and mountains. Afterward, our family drove to San Francisco along the coast, stopping at the Monterey Peninsula and taking a look at Pebble Beach.

We visited Stanford when we got to the San Francisco Bay Area, and Doug Osheroff showed us his laboratory. He told Amy she had made a good choice by going to Caltech and we all had lunch together with Doug's wife. Doug mentioned that several people in Lloyd House, where we lived, had made a bet to each buy a bottle of wine for anyone who won a Nobel Prize. He pointed out that none of them had yet paid up on the bet. (I don't think I was part of that wager.) That's when I mentioned that I was one of several people who convinced him to move from electrical engineering to physics as a major. He shot down my memory, but I'm sticking with it. Doug became head of the Department of Physics at Stanford a couple years later. Needless to say, I am very proud of my former roommate, and I told him so.

In downtown San Francisco, we got a room at the historic Fairmont Hotel. Amy was considering graduate school at Berkeley at that time, so we visited the campus. Highways had always fascinated her, and Berkeley had one of the strongest transportation engineering programs in the world. We also visited Fisherman's Wharf and the Muir Redwood Forest before one final night out on the town. That's when the trouble began.

At a downtown steak restaurant, I started choking due to my narrow esophagus. Miriam knew right away because my face always goes pale and I become silent when it happens. I tried to drink water to wash it through and then rushed to the restroom. We had just begun to eat, and I didn't return to our table for a long time, trying to barf it out in the restroom. Eventually I gave up, and we all left our food half eaten and walked back to the hotel.

It seemed like water could slip by the obstruction, so I managed to fall asleep that night with everyone worrying about me, and the next day we drove six hours back to Pasadena. By that time, even I was beginning to worry, so we all went to the Huntington Hospital and checked into emergency. A while later a doctor handed me a small piece of gristle that had lodged across my esophagus and showed me color pictures of his inside work. Later, back in Minnesota, my regular doctor, John Butler, told me I was crazy and reprimanded me for not going straight to a hospital in San Francisco. All in all, it was a very interesting family trip.

Like many people, Amy moved off campus her senior year to live in a nearby apartment with some friends. She grew a small garden and had a cat. Amy finished her classwork one trimester early and came home with a bachelor's degree in Engineering and Applied Science to start planning her wedding and married life.

Almost our full family went back to Pasadena for her official graduation in June 2001. Grandpa Ken was there along with Dan, my sister Janis, Kelsey, Miriam, and me. We would need a couple of hotel rooms for this crowd and I wanted to stay at my favorite hotel, the Huntington. When we arrived I got the bad news that our rooms had been given away to a large group and we would all be forced to stay in one room. The man at the check-in desk assured me that we would be happy, but I thought about yelling and complaining in the nice lobby. He asked me to trust him and said something about the Presidential Suite, so we let them lead us up to our single "room." Our quarters were also sometimes called the "Tournament of Roses" room. Located at the very top of the hotel on the eighth floor, it required a special access key.

We were not disappointed. It turned out that the Dalai Lama had stayed in the suite the previous week, and Bill Clinton earlier that year. It is also where all the Rose Parade princesses stay. There were always warm cookies or other treats waiting in the hall as we got off the elevator. The suite balconies extend to both sides of the hotel with a view of the mountains on one side and the downtown L.A. valley on the other. Our main floor had a grand piano, a dining room, a kitchen, and a large living room. Up a spiral staircase there were two large bedrooms and marble-walled bathrooms. Needless to say, we didn't complain about our room.

Amy had all her Caltech friends over to the Presidential Suite for a party the night before we had to move into our original two rooms.

Her graduation on the lawn in front of Beckman Auditorium was hot but celebratory. The main speaker was Gordon Moore, the founder of Intel Corporation. He was also famous for Moore's Law, which predicts a doubling of electronic device density on chips every two years. That year, 2001, also happened to be the year that Gordon and Betty Moore gave a

$600 million grant to Caltech, still the single largest education award to any single university.

We had a nice outdoor lunch after the graduation ceremony and stayed to hear student concerts that evening. Soon everyone left for home in Minnesota, and Amy and Dan began planning their big wedding.

GROW Program and Vision 2005+

You might be wondering what I actually accomplished at Honeywell during my final years there. During my tenure as VP of Technology, beginning in 1993, Honeywell yearly sales increased from about $5 billion to over $9 billion by 1999. My emphasis on new product sales, I believe, had something to do with that. My salary and bonuses were growing rapidly as the company reached all its profit and growth goals. I was giving regular talks to stock analysts, describing our future direction. We were getting the reputation as a "software and systems" company and not just an aerospace and industrial product maker. Our stock price rose accordingly.

A small group of executives started meeting for lunch to talk about the company's direction and what we could do to help. This small clan included Arnie Weimerskirch (head of quality), Steve Hirshfeld (company strategy), Ed Hurd (previous head of the industrial businesses who had been banished to Minneapolis after a couple of bad profit quarters), and several others. One person always brought celery for the lunch and someone suggested our group should be called the a-celery-eaters (accelerators). By then, the top people (Bonsignore, Ferrari, and a few others) seemed to be only interested in mergers and profits and had stopped speaking to the employees. They were unaware of our small clandestine group. Giannantonio Ferrari, my new boss, moved from Honeywell Europe to replace Larry Moore as Chief Operating Officer. Our little team discussed employee programs that we could develop to nudge the company in new directions.

I especially wanted to expand on my experience with competitive internal programs. The technology center had great success with its Initiatives competition for new research ideas, and in 1994 we began a similar program for larger developments called the Home Run Programs. The Initiatives programs had total funding of about $1 million (divided among about 10 programs) and the Home Runs had about $2 million for three to five programs. Home Runs were aimed directly at new products for the divisions, not merely research ideas.

In 1997, Steve Hirshfeld and I suggested a similarly competitive program on a company-wide scale called the GROW Program. It was aimed directly at new product developments by the operating divisions and had total funding of about $6 million per year for about five programs. The beauty of competitive programs is that you can set the rules to accomplish additional goals, beyond simply making money. For example, for the GROW programs, projects where business divisions worked together were favored. As a result, some of the programs involved joint work between business units and the technology center. Business units rarely did joint projects, so the GROW programs created new synergies and sharing. In the rules, each year we suggested new technical areas or competencies that the company needed to develop. Also, to win the corporate money, most divisions put in a lot of additional funding from their own R&D funds, so there was a leveraging effect. We used the methodology I had developed for the Home Run Program to evaluate all of the GROW candidates. I had, by then, reduced a complicated "Supertree" calculation developed by Stanford Research Institute (SRI) to a simple Excel program with embedded macros. A small group of neutral evaluators, which included Ella Ramsey (Director of Stratergy), made the project selections, which then had to be approved by CEO Bonsignore and COO Ferrari.

Some of the programs we funded with Home Run and GROW monies included: Vertical Cavity Surface Emitting Lasers (VCSELs); Air Maintenance Computers (on-board aircraft health monitoring); Fiber Optic Current Sensors (a new way to measure electric current very accurately); Home PC (for home security and control on a small computer); Atrium SW (for total control of all building sensors and systems); Smart Grid (controls for the protecting and optimizing the electrical power network); Asset Management (for overseeing products in industrial facilities); Integrated Flight Management (new capabilities for Honeywell Flight Management Systems, adding weather and flight path optimization); Airport Systems Control (ideas for fully integrating the aircraft weather, systems health, and ground facilities information); Industrial Safety (techniques for preventing and responding to dangerous situations at refineries); Integrated Pharma (products and controls for the pharmaceutical industry, a new Honeywell market); Retail Restaurant Management; and System-on-a-Chip (merging functions of many computer chips onto one). Those programs were together projected to generate about $2.5 billion of new revenue by 2004, a huge return on the investment.

One program I'm especially proud of was the Vertical Cavity Surface Emitting Laser work, which I supported as an Initiative, Home Run, and then GROW program. Using VCSELs, two chips placed over one another can communicate directly through thousands of light beams rather than by wires at their edges. There are numerous other applications as well. By 2000, Honeywell VCSELs already had become a $30 million per year business with growth to $100 million a couple years later. VCSEL products from Honeywell and others helped to fuel the growth of the internet by becoming the laser source of choice for Local Area Networks (LANs) and Storage Area Networks (SANs). The key researcher throughout this time was Mary Hibbs-Brenner, a Stanford PhD who is now CEO of a VCSEL company called Vixar in Plymouth, Minnesota. I especially wanted to use the GROW program to continue this development in the late 1990s because the technology center had run out of funding.

Mary, of course, won a Sweatt Award for her work and I especially remember one extra reward she received. She had gotten married during a post-doc in Sweden, and both Mary and her husband were invited by HTC to a Timberwolves basketball game one evening after winning the Sweatt Award. It was the only time I have actually been seated in the front row, right next to the team. We could watch the players sweat. They towered over us, especially Randy Breuer, at 7 feet, 3 inches tall. Mary's husband was completely astounded by this display of American culture, and I remember him watching bug-eyed during the game. I have been a Timberwolves fan ever since, despite the fact that they can't seem to win, at least so far.

By 1998 the idea of competitive company-wide development had spread and Honeywell started a program called Vision 2005+. The goal for those programs was to create billion dollar revenues in new market areas by 2005. Steve Hirshfeld led this effort, which pulled 20 of the most respected young "fast-track" employees from their divisions to develop the new businesses. They expected to be separated from their normal business units for two years at least. I was one of the evaluators as the new areas were identified and selected from many candidates. By 1999 three businesses had been chosen: Aviation Enterprise (essentially an airport/aircraft management business), Bio-Pharma (an opportunity to use Honeywell control products and systems in the growing pharmaceutical area), and Asset Management (an enterprise-wide management system, including logistics, for industrial businesses). Each program received $2 million development funds in 1999 and were scheduled to get $5 million in 2000. The management structure and leaders for each of them were beginning to be selected by the middle of 1999. I mention that because, as will be explained later, I wanted to become part of the Aviation Enterprise business. Unfortunately, by 2000 all of them had essentially been put on hold or killed.

These competitive programs have shaped my views more generally about both economics and politics. My early studies, back in my first manager position, had proved to me that competition is a great way to speed up development and creativity. I think that is partly why capitalism seems so much more effective than communism, for example. Competition in business, sports, and many other areas spurs improvement. On the other hand, many people, especially Republicans, believe that letting everyone keep their money and make decisions on the smallest scale possible is best. They believe that competition will naturally lead to the ideal use of resources. But people don't naturally join up to build roads, keep the air clean, or work together in siloed corporations to start totally new businesses. It takes top-down leadership, or inspired entrepreneurship, to go in new directions and to make needed changes. We require an active government and courageous company leaders to set new directions occasionally. That's why I felt so strongly about our Competitive Development programs at Honeywell—they combined a top-down nudge and funding within a competitive structure to bring out great creativity and success.

Tornado

Late Friday afternoon, May 15, 1998, Kelsey was home from school thinking ahead to her confirmation party. Dozens of nearby friends and relatives were expected to attend a huge party at our house Sunday afternoon. The temperature and dew point were unseasonably high for May. It was a clear day, but threatening clouds were moving in from the west.

By 4 p.m. heavy rain had started and Miriam was getting ready to drive home from a haircut, about five miles.

As the storm worsened, I walked through the halls at the HTC Camden building telling everyone to stay away from the windows and not to go out in the traffic. Tornado warnings were spreading. I literally shouted, "Keep working. It's safer here than on the roads." Looking out the windows, the sky was deep black with high winds, and people began to move toward the interior of the building. Around 5:30 the storm seemed to be letting up, and I got a call from Miriam saying I needed to get home, that our garage wall was half missing. Normally the drive home took only 20 minutes. That day, avoiding flooded freeway underpasses and fallen trees, it took two hours.

Kelsey had been hiding in the basement when she heard windows breaking. She was alone and scared in our storage area as the tornado roared over our house. Miriam drove home during the worst possible time and her minivan rear window exploded out, probably from the tornado passing directly overhead. Roads were impassable later from numerous fallen trees when I tried to make it home. By the time I got close to the house, police were everywhere. I had to prove I lived in the area to be allowed passage.

When I finally reached home, I saw the garage door was ripped off, one side wall of the garage was dangling half open, and windows were broken everywhere. The power was out, but luckily, the gas line on the garage wall didn't break and held the wall from falling. The twelve 50-foot cottonwood trees near our pond were broken about 20 feet from the ground as if the tornado had grabbed the top part and twisted, with the bark ripped sideways. A large tree in our front yard was pulled to the ground and a tall evergreen was shoved over about 20 degrees and littered with small items from the garage. We later learned that the tornado touched down in Roseville, and then traveled 12 miles through Shoreview, North Oaks, and Lino Lakes before lifting back up into the clouds in Blaine. Seven houses were destroyed and several hundred more were damaged. Total damage in our area was about $151 million. Some houses right around us had far more significant damage. One was lifted up off its foundation and then set down at a slight angle. Our neighbors across the pond had their entire large porch ripped off and deposited in the middle of the street, with the furniture and TV still in place. Almost everyone had roof shingles removed, which flew around the area breaking windows. We found shingle pieces in our bushes two years later. Trampolines and yard furniture flew miles away or were deposited in our pond.

Our home damage cost $40,000 to repair. We were lucky compared to many neighbors. For weeks, fallen trees were being cut up and placed near the streets for the city to haul off. Almost everyone had siding replaced and planted new trees. A year later, it looked like a brand new neighborhood.

We moved fast and were one of the first to get insurance money and begin repairs, which still took the entire summer to complete. The best result for us was new carpeting for the basement. One small basement window had been shattered with glass pieces falling onto the blue carpet there. An insurance rule requires that carpets be replaced if even a little glass might be imbedded, which for us meant that carpeting in three connected basement rooms had to be completely replaced. This was great because the old one had epoxy stains and glue everywhere from building the yellow submarine. Even more, the carpet company no longer made the color we had used, so they had to do a special dye lot to make our new basement rug.

We had landscapers redo our entire yard, planting birch trees and building a fire pit for Kelsey. I hitched a block-and-tackle line to the tilted evergreen and tugged on it for three years. It is now growing vertically.

Most important, Kelsey's confirmation ceremony and party were a great success. We simply moved the party to our church building and everyone had a good time while Miriam and I worried about our house and the food warming in our unpowered refrigerator and freezer.

The Day It All Came Crashing Down

My life as a Honeywell exec was flowing along pretty well by early 1999. When I didn't have six meetings every day at the office, I was flying around the country or the world for more meetings. As mentioned, I now had three offices, including one at the corporate headquarters. Karen Bachman, Honeywell's head of public relations, badly wanted an office on the 5th floor, so I was asked to vacate my seldom-used office. But to compensate, the corner office suite, normally reserved for the head of Honeywell's overseas operations, was given to me. It was actually a four-room suite, with two conference rooms, my large office, and a huge entrance area with three desks for secretaries or assistants. There were international paintings, sculptures, and time clocks everywhere, and I had a beautiful view of downtown Minneapolis. My secretary Carol and I probably only actually spent 10 days in that wonderful office, favoring our space in the Camden R&D building.

I took two interesting trips around then. I had been asked to give a talk in a Paris suburb for the European directors. To avoid getting jet-lagged, I decided to stay on Minnesota hours as much as possible, so I slept on the plane, took a train from Charles de Gaulle Airport, and went directly to sleep in the hotel where the meeting was to occur. My talk was scheduled for 9 a.m. in France, or about 3 a.m. Minnesota time. So I awoke, gave the talk, and went directly back to bed until I needed to get to the airport for the flight back. It worked; I wasn't jet-lagged but did get a headache.

The other crazy trip was to the Grandview Lodge on Gull Lake in northern Minnesota. All Honeywell lawyers were meeting and wanted me to tell them which technologies most needed patent protection. They apparently had a large budget because they asked me to fly alone on the company's small jet. I remember taking off from the Twin Cities airport, going up at about a 45-degree angle, and seeing Lake Mille Lacs only about four minutes later. The entire trip took about 20 minutes, most of which was lining up for the landing. A car was waiting at the Brainerd airport. The next morning I flew back, alone again, and was able to enjoy a beautiful view of the fall leaves blanketing the Twin Cities.

When Ben Simmons was offered a new job leading an inertial navigation business in Florida, I needed to hire a new Vice President for the technology center. Rather than promoting from within, I felt that HTC needed a new viewpoint, so we hired an external search firm. After several months, we found an amazing research leader from 3M, Dr. Kris Burhardt, for the job. Kris's life story was full of both technical and personal accomplishments, which included escaping through the border of the Iron Curtain from Poland. His selection was a surprise for HTC, especially for those competing for the job.

Everything was not perfect. Seven years on the same job, no matter how wonderful, is a little long. I think three to five years is optimal, both for the person and for the group. An MIT study found that four years between shakeups were best for successful teams. I was getting antsy. And a few events made me nervous. At the strategic planning meetings with top management, I had introduced a new company technology direction: Information and Knowledge Management. With the advent of Google and similar internet capabilities, the understanding and protection of a company's key intellectual property had become more important. I had started to analyze each of Honeywell's main business lines and presented the results at the General Managers Meeting.

The company needed to think about its core proprietary information. But my talk seemed to go over the head of Michael Bonsignore. He appeared more concerned about protecting existing product businesses. He was also upset when the military avionics business asked him to buy a new technology, Gyro on a Chip, for $25 million. This new device was being developed at MIT's Draper Labs. Most of the technical leaders thought this a long-range research subject and premature to spend a large portion of the budget on. We were already developing RLGs, and fiber optic gyros, but the panicky business division won out, probably because our main competitor, Litton, was after this technology. Bonsignore was unhappy having to lay out so much cash but ultimately couldn't resist.

In May 1999, Michael Bonsignore dropped his bombshell. Honeywell had decided to merge with Allied Signal, a company about twice the size of Honeywell located in New Jersey. Bonsignore had been in secret talks with Allied's CEO, Larry Bossidy, off and on for a couple of years. Only a few Honeywell people were in on the final negotiations. Honeywell's name would be retained but the headquarters would move to Morristown, New Jersey. Over 500 people at Honeywell's corporate headquarters were to be laid off and a 115-year tradition of Honeywell in Minnesota would come to an end. Bonsignore would become CEO and Bossidy Chairman of the Board until his nearing retirement.

My boss, Giannantonio Ferrari, soon told me that I could retain my job, title, and salary, and not have to move to New Jersey. But it was clear that to continue with the new Honeywell International, I would be flying to NJ a lot. When I asked Miriam about moving there, she said, "Fine, but Kelsey and I are staying in Minnesota." I had a contract that guaranteed a two-year severance package, so a period of agonizing about what to do began.

Around July, the Allied people came to Minnesota to tell us about the new company. They talked about their six-sigma quality programs and listened to our people describe their businesses. But the overwhelming impression for me was the apparent joy their spokespeople had about laying off employees. Everything at Allied Signal was about reducing costs. My goals had always been about retaining excellent employees and building great, valuable, exciting new products. I was disgusted by their obvious happiness about ending people's livelihoods.

Around the laboratories, everyone wanted to know their fate. The labs would be retained but broken into pieces and dedicated to specific businesses, thereby destroying the systems synergy that had been one of Honeywell's great strengths.

In August, I went to a decision meeting with my boss, Giannantonio Ferrari. Being very loyal, I was looking for any word of encouragement from him about the future of the combined companies. Even he, the Honeywell COO, had only been brought into the merger negotiations in the final days before the announcement. He said he would be moving to New York, but was very uncertain if that was the right move for him. Instead of encouragement, all I heard were reasons not to stay with the new company.

Ultimately, there were four reasons why I decided to leave Honeywell:

1. I was getting tired of the grind. My health was deteriorating and I was itching to work in a smaller company where innovation wasn't so slow and difficult. Flying regularly to Morristown, New Jersey, wasn't very appealing.

2. If I kept working for Honeywell, the special two-year severance package they were offering might go away. That package was based on my average compensation during the previous five years, which had been substantial, including salary, restricted stock, and bonuses.

3. Allied Signal was a boring old-product manufacturer with a mid-1980s mentality. Everyone could name a dozen companies

that were better merger candidates, with better technologies and businesses. And Allied seemed to love eliminating employees.

4. At my meeting with Giannantonio, I was told that the Vision 2005+ programs were being put on hold. Each would receive about half a million dollars to keep the teams together, but nothing like the $5 million for 2000 that they expected and needed. The new company was, at least temporarily, pulling the plug on new innovations.

For all those reasons, I told the company to lay me off (just quitting would have voided the severance package). My restricted stock vested (meaning I could sell anytime) and my payday upon leaving (including golden parachute tax compensation) would be astronomical. By that time everyone in the company seemed to only care about the merger details. The businesses were suffering.

That Christmas, rather than having the top Honeywell people over to Bonsignore's skyscraper condo in downtown Minneapolis as usual, we all gathered in a quiet restaurant room. His first words to all of us, most of whom would be leaving the new company, were "Don't be maudlin tonight." In Bonsignore's eyes it was all "just business," that everyone would bounce back. He would be guaranteed about $30 million in the deal and would live in Manhattan with a helicopter ride to Morristown each morning for his CEO job.

The Honeywell Chief Counsel, our top lawyer, hadn't even been informed of the merger until the last minute. At our table, he was clearly bitter, incensed by the whole thing, and mumbled curses he wanted to say to Bonsignore. None of us were happy, but the evening was fairly civil.

The kindest interpretation of the Bonsignore decision was that Allied Signal threatened to buy Honeywell anyway and that he avoided a hostile takeover and negotiated the best possible deal. The least-kind interpretation would be that Bonsignore simply sold out Honeywell due to personal greed. My guess is that the reality is somewhere in between, that he needed to do something bold to show for his time as CEO. Most executives believe that they are generally right.

I had been promoted eight times to my final position. It happened because

I did fearless, successful things along the way. Michael Bonsignore had been CEO for seven years and probably felt it was time to do something big.

In the end, only about seven people moved to the East Coast from Honeywell, and within a couple of years almost all had left, including Bonsignore. Giannantonio moved to New York City and had a limo ride to work each morning. My friend Steve Hirshfeld moved to the city as well but had to drive himself.

How Bonsignore himself got booted is interesting. He tried another merger with United Technologies (UTC), but it got scuttled at the last minute when General Electric sent in a larger offer to Honeywell's Board of Directors the final weekend before the expected UTC merger announcement. Several months later, the European Union outlawed the Honeywell merger with GE because it would create monopolistic businesses, especially in aerospace. A few months later, Michael Bonsignore was kicked out, and Larry Bossidy was brought back to run Honeywell.

On January 31, 2000, Carol Warne and I left the HTC Camden building for the final time. By company rules, Carol's job had to be eliminated when I was laid off. Chris Burhardt took over as Chief Technology Officer for a short time before he also left the company.

All in all, I'm really glad to have left the company early and not participated in two years of idiotic maneuvering that had nothing to do with creating real value for the world.

So ended the primary working period of my life. I had studied for 27 years and worked for 27 years. It was time to move on to Part III.

Engagement announcement, April 1973

Wedding, September 22, 1973

First Honeywell project—a solar-absorbing sphere.

Solar coating tanks, with Henry Mar.

Backyard of our new home, 1977

Miriam, ready for a baby.

Ron wins Sweatt Award,
Honeywell's highest technical
honor, plus a free trip to Europe.

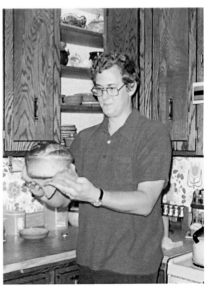

First successful raspberry soufflé.

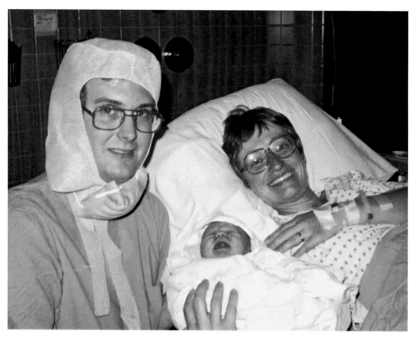

Amy is born, September 1, 1978

Amy at nine weeks old.

Ron's father, Ed's, last birthday, 1981

Ron is promoted to manager, July 1982.
Dr. Quack rules!

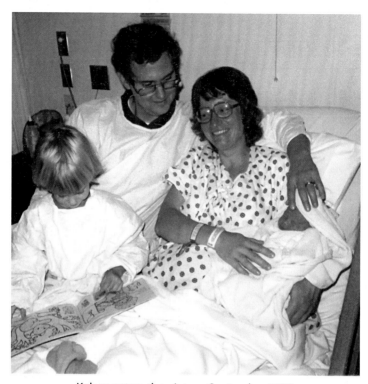

Kelsey enters the picture, September 1982

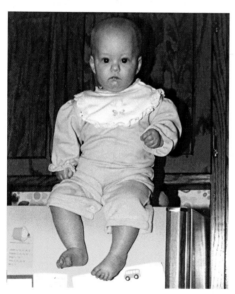

Refrigerator jump, a family tradition.

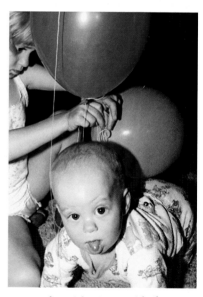

Amy tries to get rid of
her new sister.

Bedtime reading.

At Island View Lodge toga party, 1985

Kelsey becomes spunky, June 1985

Rabbit visit on Easter and
Amy breaks her arm, 1987

Typical Halloween, 1988

Ron's promotion to Vice President, April 1988. Celebration at Janis's house.

Golfing with Ido-san on the second trip to Japan, 1991,
when Honeywell sued Minolta.

Miriam and Ron head off to the Honeywell Board of Directors meeting in Sea Island, Georgia, 1994. Limo and jet provided by Honeywell.

The family goes to England, 1993

Standing on Hadrian's Wall.

Ron's 50th birthday party.

India trip, Ron at Taj Mahal.

Beginning of the Honeywell India Software Operations in 1996.
Now with over 9,000 employees.

The Minnesota Technology Corridor Corporation Board of Directors.

Ron judges the European Futurist Competition in Munich, 1997

Submarine launch day, September 1996

Local press and TV coverage.

Amy's official H. S. graduation picture.

Sweatt Award Gala, with America's Cup sail in San Diego.

Stanford visit with Doug Osheroff.

Tornado hits, May 15, 1998

Tornado damage in our backyard.

PART III

Giving Years

Moving On, Moving Mom

There was no master plan as I moved on to stage three of my life. I foresaw no dangers, only opportunities. A PhD scientist with 27 years of successful work, I now had the freedom to start over or to do nothing. The dot-com bubble was at its peak, our investments were soaring, I would soon receive my Honeywell pension plus the two-year severance package. In a few years Miriam would have her pension, and still later, Social Security would kick in. Money was not an issue. On February 1, 2000, I did not drive to work at Honeywell as usual. Avoiding rush hours, I soon learned, is one of the greatest joys of retirement, or whatever I was now doing.

In January my mother signed official papers for a new condominium in Summerplace of Shoreview. She would be only a few blocks from our house and a couple miles from my sister Janis. This decision had been delayed by nine months after a fire (arson) struck during construction. She had lived in her house on Chamber Street in St. Paul since 1960 and saved all the papers she found interesting or important in boxes in her basement. Soon I started hanging out there, sorting out what seemed useful. My mother was old school and only trusted me (the man) to actually throw away any papers, especially financial or medical records. Her entire history was mixed together: medical receipts with old newspaper articles, photographs, memorabilia, clothes, etc. Most was toss-worthy, but from time to time a priceless photo, award, child's drawing, or gold fillings from my father's teeth were discovered in the mix, which meant that every item had to be examined. It took about a month just to go through the papers.

In April we contacted a real estate agent who said her house was worth about $110,000, but more if it was fixed up, so we sanded and resealed her hardwood floors after removing the carpets. Taking the carpets out was the hard part since the under layer of foam had bonded with the flooring and had to be hand-scraped. When finished, it looked great.

We held a massive garage (estate) sale that netted about $1,500. The first day we sold lots of items (my dad's hand tools were a particularly hot seller). By noon of the third day we had reduced the prices dramatically, and by 4 p.m., anyone looking at the items had to promise to carry something away for free.

Her house sold for $129,900 because the housing market had started to improve and the place looked a lot nicer after our work. She moved on June 27 into a two-bedroom condo located across from the Shoreview Library, community center, and park. She was 87 years old and we could easily walk to visit her. This third phase of my life would be primarily to help others, and my mother was the first.

Life without official work never seemed boring. I started calling myself "semi-retired." There were trips with Kelsey to check out colleges and later to visit her at Valparaiso. I became a member of the Galilee Church council and continued singing tenor or bass with the choir (whichever was needed more). Filming Kelsey's sporting events became a new hobby.

I put in some working capital and formed a consulting company with a lawyer's help, called Peterson Technical-Business Strategies, or PTBS for short. It was an S-Corp and I even had a fancy intricate stock certificate. I became the owner, CEO, COO, CTO, and janitor of the company. As I will describe, clients trickled in over time without actively marketing. I was busy, but there was no pattern to it yet.

Around March 2000, the dot-com bubble burst and the stock market began its great fall. I decided that keeping a close watch on our existing finances was more important than earning new money. After leaving Honeywell, about one-third of our investments were with a company led by Mike Ovshak. Starting in the 1980s, we had been saving money with his (or his father-in-law's) company regularly. With the Honeywell options, stock, and severance, a lot of new money was available and we invested that with a different company led by Brad Johnston of the Johnston Group. Unfortunately, Brad had shifted some of our money to the soaring dot-com and high-tech companies on the NASDAQ exchange just before the crash, so we lost a lot but not enough to become concerned. The NASDAQ dropped from 5,000 to only 1,000 a couple years later, and has only now (2018) recovered to 7,300.

I did pay more attention to our investments after that, making elaborate tables and graphs, especially at tax time. Over the years, both investment advisors have done well for us and become our friends. Having two counselors also tends to keep them more alert.

Overall, my life became a series of projects, some short-term and some continuing. I could sleep late and wear comfortable old clothes almost all the time. With Miriam still working, my time was fun and quiet.

A Job?

By the summer of 2000, various job opportunities started to appear. Most of my Honeywell colleagues had quickly gone into full-time job search mode, but that had not been my priority. Many of the Honeywell executives were managers, bureaucrats who needed employees to do their work. That wasn't me. But some exciting options were popping up, so I needed a rule. I decided to apply only for jobs that had a chance of changing the world, or at least a chunk of it. I would not apply just to have work and make money. Eventually, I applied for two jobs.

The most exciting was an opening at the University of Minnesota for a new organization called the Digital Technology Center. The state legislature had provided new funding for hiring about 15 new professors to focus on advanced digital devices, wireless, biology, medical, and genomic research, a cross-functional attempt to jump-start research and new companies in Minnesota. The governor, legislature, and University Board of Regents had all supported this fledgling organization. Support would come from many sources and would eventually require new laboratories in the U of M's Walter Library. About 200 people applied to be the director of this new organization, and the Dean of the Institute of Technology (IT) at the university urged me to apply. I knew Dean Ted Davis from my work on his advisory board. So my name went into a large pool and I waited for about 10 months.

Meanwhile, I interviewed for the chief technology position at a company called ADC Telecommunications in Plymouth, Minnesota. ADC made devices for the phone industry, many of the switches and panels used at the base stations. They were intending to move into wireless systems and fiber optic lines from the base stations to homes (the so-called last mile of the system). This was a very exciting company that had been growing rapidly, so I went through the interview process and came close to getting the job. At the last minute they decided to promote someone with years of experience at the company and moved away from their new goals. They had tried to start a new software business, had a few bad quarters financially, and became hesitant to do new things. During the 2008 recession it became even more difficult, and in 2010 Tyco Electronics purchased them.

By 2001, the Digital Technology Center opportunity finally became real. I started studying all the technologies that might be needed for the job, especially biology. I had never taken a biology course at Caltech, and it wasn't one of Honeywell's focus areas, so I was about 40 years behind. Amy loaned her Caltech book called *Molecular Biology of the Cell* to me, and I started reading the 1.5-inch tome. Ted Davis called to say I was one of the two finalists for the director slot, and we set up a series of interviews with the eight to 10 professors who had already joined the group. This was a wonderful experience. I had great discussions with advanced computer researchers, biologists, wireless experts, and medical scientists, all of whom were working cross-discipline. There were lunches and a fun final dinner with about seven professors and department heads at a beautiful restaurant.

My final presentation, with about 30 university people attending, was where I probably lost the job. After describing my history and research, which was mostly done in the 1970s, I talked about how the new Digital Technology Center could be a huge competitive advantage to Minnesota. I had used my Honeywell information analysis tool to identify areas of great importance where discoveries would best position Minnesota for the future. One gnarly math professor on the selection committee was quite upset by this. He said, "We believe that all knowledge generated at the university should be for all the world, not just Minnesota." I'm not sure that's the way the governor would have described the reason for millions of dollars of new spending, but it probably is the way the university thinks. And of course I believe that as well, but it is critical to focus any research using some criteria, not just letting it run haphazardly.

Ted Davis later told me privately that my interviews were the strongest, that I had the broadest background, and that he personally thought I was the best choice. But the job went to Andrew Odlyzko, a Bell Labs director, who had continued to generate research publications during his management years, which I had not. He had been doing research on how internet growth had been overestimated. I hadn't published enough. The DTC Director position would also have a full professorship title attached, and I would have to get back to doing physics research quickly. This seemed like something I could do, but it probably was a concern for the selection committee. Andrew is long gone, but over 40 university scientists and professors are connected to the DTC. It would have been a fun job.

I mention my failure to win the DTC position to show that not everything has gone perfectly in my life. Most successes were balanced with struggles. Perhaps it was okay to not be selected. I would have gone back to working 80 to 100 hours per week; it's difficult for me to not throw myself headlong into new challenges. I might have had a heart attack by now, rather than writing this book.

There were still plenty of things to do in my "semi-retirement."

Amy and Dan

As mentioned before, in 2001 Amy finished her classes at Caltech a few months early and drove back to Minnesota with Dan to start planning their wedding. Dan was a physics junior at the University of Minnesota at the time, and they planned to live in student housing after their August wedding. Four wedding showers also filled Amy's time—college friends, Minnesota church friends, Minnesota relatives, and high school friends. Amy is an outstanding organizer, so she juggled the many balls well.

The actual wedding occurred on a stifling hot August 4 day. Dan has numerous uncles who are ministers and one flew in from California to officiate at the United Methodist service in Golden Valley, Minnesota. We all blew soap bubbles as they hurried out of the church, and then attended a reception at Blaisdell Manor in Minneapolis. I had a video camera set up where guests could offer their thoughts and encouraging words to the new couple. Dan's father, Will, was funny. He wandered around the far field of the camera for about 10 minutes before finally stepping up and reciting a beautiful short poem he had just composed to encourage the new pair.

After opening presents at our house, Dan and Amy drove off for a vacation in Door County, Wisconsin. They soon moved into a university apartment before Dan's senior year began. They had everyone over for a celebration dinner in late August. The hallways there were filled with the aromas of cooking from many countries. Their super-long engagement was finally over.

A couple weeks later, terrorists struck the World Trade Center in New York and the Pentagon in Washington, D.C. Miriam and I watched the whole disaster on TV. I don't think anyone understood the full consequences of that day at the time. The entire country and world entered a new mode of operation, which is probably a lasting change. There was a weird mixture of ordinary life, which President George Bush encouraged at the time, with undercurrents of fear. It reminded me especially of the time when we led normal lives but built bomb shelters and practiced for a nuclear attack during our school classes.

Anyway, that's the environment in which Dan and Amy began their married life. Dan is a thoughtful, serious young man, whom I like very much. I never understood how much Amy admired him back when the yellow submarine was under construction, my being oblivious to such things, but it all worked out well. After all, who couldn't like a fellow physics person who also plays piano and is passionate about music. At his graduation we purchased an electric piano as a present. It is a joy to hear him playing at family gatherings. Dan finished his senior year and graduated from the University of Minnesota the following spring.

Amy settled into married life, and in December 2001 got a job working for the Minnesota Department of Employment and Economic Security (the people who define the budget available for the state legislature each year). She became a Research Analyst doing various studies, including the primary analysis of the new census data, making new maps for Minnesota, etc. A few years later, Amy created her own web company called Ewe Betcha (being part Swedish and Minnesotan) and sold knitted items, especially woolen diaper covers for babies. I particularly liked the ones with skull-and-crossbones on the bottom. The company did very well, especially at selling fine yarn that she bought directly from a farm in Wisconsin. She also has done research and evaluation on thousands of published articles for several companies, part-time from home.

At Christmas 2002, Dan surprised us all at a gathering by saying that he had decided to become a Catholic. He explained that most religions were just too arbitrary and he wanted a faith that had a long history and clarity about its positions. His Methodist minister, who talked about God being a She, particularly bothered him.

While I could understand his reasoning, I must say that the decision bothered me. My whole background, especially my years at Caltech, had taught me not to blindly accept the pronouncements of "authority figures." And in my opinion, the Pope, being human, is demonstrably not flawless. I also considered the Catholic position that they were the one true religion to be offensive. This was reinforced when Dan and Amy had to redo their confirmations to be official in their new church. But there are far worse disagreements in most families to worry about. I would be truly disappointed if they didn't build their lives together in the best way they saw fit.

Dan worked awhile for Promet, a company that made optical thumbprint detectors, took teaching classes at Hamline University, and in September 2005 became a high school physics teacher at Providence Academy, a Catholic private school. His father, Will, had been an English teacher at a private school during his 40-plus year career. Dan's school gave him greater flexibility to build programs to reinforce the academic learning for students of all abilities. To update Dan's latest work status, in 2016 he began working for OSI (Open Systems International, Inc.) as a Power Systems Engineer.

Amy and Dan rented an upstairs apartment in the St. Paul Midway district in November 2002, and later bought a house on 37th Avenue in Minneapolis. In June 2003, at a party for my birthday and Father's Day, they had a big announcement. That's a story for later in this book.

More Cool Trips

Miriam and I had some fascinating travel experiences in the early 2000s. The first took Miriam to South America with some of her good friends in August 2000, visiting Argentina, Brazil, Chile, and Peru. She traveled with Lynda Thompson and Judy Reuss and they met up with Meggie Thompson (one of Amy's friends and Lynda's daughter) who was studying Spanish in Argentina. Meggie's out-of-country experiences paid off; she became a finalist for Minnesota Teacher of the Year. The group saw Iguazú Falls, ate at an Argentine steak house, sunbathed on Rio beaches, flew over the Andes, and had a generally great time.

I missed all that, but joined up with Miriam, Lynda, and Meggie in Lima for the Peru leg of the trip. Flying up to Cuzco, which is at an altitude of over 11,000 feet, had an interesting effect on Miriam. Ignoring advice, we immediately went out to visit the Incan ruins there, and she fell ill from altitude sickness. Miriam never gets sick, but she was vomiting and had a terrible headache and more. So she got a ride back to our hotel while Lynda, Meggie, and I finished the tour.

She recovered just before we headed off on a train to Machu Picchu. Up on that mountain we talked to the llamas, climbed a short way up terrifying Inca stairs, had a guided tour, and stayed at the only small hotel at the top. The next morning, before sunrise, we walked out on the ruins. Misty clouds were rolling below us on three sides. No other tourists were present. As the sun rose, it glowed a golden color through the fog blowing gently over us, and the timeless sense of the people who lived there over 500 years earlier affected us all. Both the amazing stonework and the breathtaking location have burned images into my memory. The photographs I took of this World Heritage site help too, of course.

Part of that South America trip included Judy Reuss, whom we've known for about 20 years. Judy and her husband Bob had sold their home in Minnesota, boarded a boat in Florida, and sailed the Gulf and Atlantic coast for over four years. We think Judy got tired of the sailing life after awhile and they settled down in North Carolina. We visited them in 2002 and first saw all the civilized places there: Duke University campus, Raleigh/Durham, the Biltmore Estates (built by George Vanderbilt, it's the largest private house in America), the pottery factories near Asheville, and the Blue Ridge Parkway.

One night, staying at a motel along a mountain creek the night before the big fishing opener, we had a harrowing experience. Overlooking the creek, dozens of people were camping out below our motel room, partying and making a lot of noise. I was reading quietly in our group of rooms when a string of gunshots rang out. Running to the window, I saw a pickup, wheels screeching, accelerate up the dirt road and out of sight. I realized then that opening the shades with bright lights behind me wasn't a great idea. A minute later, someone pounded on our door, screaming to come in and use the phone to call an ambulance and police. This was a difficult moment—do I help the man with someone injured but possibly risk getting shot ourselves? I yelled back, "We don't have a phone in here. Go talk to the motel manager." Miriam said that was true, we didn't actually have a phone in our room, but I still feel a little guilty about not helping those crazy people.

The police soon arrived. They had caught the shooter, the "Grandpa" of the clan by the creek. He had been speeding at over 100 mph down the road and the police pulled him over. We learned the next morning, as the police crawled around searching for shell casings in the dirt, that Grandpa had fired five warning shots over the head of his daughter's boyfriend, but winged him a bit. Apparently he wanted the boyfriend to make his daughter an honest woman. I'm pretty sure a shotgun wedding was in their future.

In 2002, Miriam sucked me into a Spanish student trip to Costa Rica. I served as videographer for the group, which would become one of my recurring gigs. In fact, I have made about two dozen short movies of weddings and other events since then, getting paid occasionally. In Costa Rica, I made movies of crocodiles, waterfalls, volcanoes, rain forest sloths, monkeys, and hummingbirds with their wings slowed down. A highlight was up in the mountains, near the Arenal volcano, where everyone tried the zip lines. I made movies holding the video camera in my mouth, zooming down 200 feet over the ground.

The zip lines were where I learned what terrible shape I was in. I had slowed down the whole group climbing up hill after hill, and then ascending a forest ranger tower at the highest point. Stopping frequently to catch my breath, Miriam and I worried about a possible heart attack.

I was 57 then, about the age when my father had his first attack and I weighed about 230 pounds, in the slightly obese range. Thus began my 15-year battle with weight control. This had an exclamation point later back home when, during a routine stretch of my esophagus, I had a 30-second mini-stroke while coming out of the procedure, losing feeling in half of my body. This has never recurred, but it caused a major scare and an ambulance ride to the hospital. For the next several years I took regular hour-long walks and monitored my steps on a Smart Brain pedometer connected to the internet. I have five years of charts to look at if you are interested. My weight stabilized but didn't drop until about 15 years later when I found a way to miraculously lose 40 pounds. My secret formula: eat only nuts for lunch, cut back on the Coffee-mate, eat lots of fruit and vegetables, and generally exercise "portion control."

Another trip in 2002 is worth a mention. Miriam and I drove alone to the Berwood Hills Bed and Breakfast. This gorgeous old farmhouse and grounds near Lanesboro in southern Minnesota was packed with antiques and quirky history from the previous 100 years. They served gourmet meals to the small number of guests there and it was a perfect place to escape the routine and spend a romantic weekend. Aah!

Galilee and ISAIAH

Although I was getting occasional consulting jobs, to be described later, in the early 2000s my time was flexible. I could easily volunteer, and it also meant people felt open to ask for my help. I began to utter the phrase, "Competence is a curse!" fairly often. Two regular volunteer organizations were my church, Galilee Lutheran, and a social action group called ISAIAH.

I had attended Galilee, taught Sunday school, participated in the rituals, and sung in the choir since the late 1970s. I will describe my slightly odd relationship to God and religion later, but commitment to this small, outward-looking, multicultural church was never an issue. In 2000 I served as co-chair of the church's 40th anniversary. Our celebration began with speeches at the neighborhood elementary school, where the congregation got its start, and then parading over to our current building. The bishop of the national church, the Evangelical Lutheran Church in America (ELCA), was on hand and gave a sermon, during which a few birds got into the sanctuary. His name was Mark Hanson, a man Miriam knew at Augsburg College, and he said something about the Holy Spirit before the birds finally flew out again. Afterward, I played my 30-minute movie and slideshow about the 40-year history that everyone watched, and still later, a time capsule was buried and balloons let free.

I had been on the congregation council twice before, and in 2003 became the council's President. Applying my management and creativity principles, the congregation started some new activities. We set goals: "Reaching Out," "Growing Closer to God," and "Everyone Involved and Committed." A "Vision to Action" team was established to make things happen.

As a result, the entire congregation spent Sunday evenings together in small (six- to eight-person) groups in homes for six weeks to explore their lives and faith. One elderly woman, Lois LaVon Brase, tripped on our sunken living room step, fell into our piano, and crashed to the floor. She said not to worry, and we continued the meeting for 30 minutes before she got off the floor.

Everyone read the Bible chapters of Matthew together and discussed each week's readings while eating bread and soup during Lent.

Perhaps the most memorable new activity was when we invited famous groups or singers to the church for interesting evenings. One night we invited the notorious Lutheran Ladies for a night of comedy. They were famous for mocking normal Lutheran stereotypes. The two women were also friends Miriam knew from her time at Augsburg College. Part of their act involved making fun of Jell-O, a typical Minnesota Lutheran dessert. So I insisted that we have a Jell-O contest as part of their visit—and I intended to win.

I worked on my Jell-O creation for two days, using a 4-inch diameter PVC pipe as my mold. I bought every known flavor and color of Jell-O and carefully froze one after another into the 28-inch pipe. A special plunger was designed to push the full tube of Jell-O out after heating the pipe to make it slippery, creating a long Jell-O rainbow log roll. At least that was the plan.

On the comedy night, the whole church was packed, and after the Lutheran Ladies act they called for the Jell-O creations to be brought to the front. There were about 10 contestants, including an American Flag Jell-O design that looked great. But it all happened too fast, and I didn't have time to heat my PVC pipe. So I had to push my Jell-O rainbow out onto a large tray too quickly. The colors got all mushed together and it all plopped onto the tray in weird wiggly formations. The Lutheran Ladies looked at me in disbelief. I then experienced a personal comedy barrage, complete humiliation, and rolling waves of laughter from the audience. They gave me some kind of booby prize for worst dessert. Moral of this story: don't try to make your Jell-O in a plumbing pipe.

That was also the year we started inviting homeless people into the church during December. Beds were set up by an organization called Project Home, and we fed families, stayed overnight with them, and generally got to know and understand their problems. Primarily we learned that people are people, and that any of us could fall into a homeless, hopeless situation. They had arrived at our doorstep due to health issues, loss of jobs, and problems with cars, food, and money. Helping their children was a joy.

Our pastor, Dick Carlson, was enthusiastic about a social-justice organization called ISAIAH and encouraged me to start going to their meetings. About 10 churches were involved in the Northeast St. Paul region. Each subgroup of the larger 100-church organization got to vote on their top issues, and we selected education. This was rolled together in a large meeting of all the churches and became one of four ISAIAH focus issues that year.

ISAIAH is an interesting organization in that it doesn't directly provide charity assistance. The goal is to work upstream, to lobby or persuade government, powerful organizations, or businesses to fix the root causes. They do this by a variety of means, one-on-one relationship building, group meetings, and creating community visibility on key issues.

In 2002, Pastor Carlson convinced me to go to Chicago for ISAIAH leadership training. It seemed unnecessary for me after I'd been through at least 10 weeks of leadership training at Honeywell. But off I went to Chicago with a local pastor and a man from St. Cloud, Minnesota. The Gamaliel Organization provided the training, the same training group that convinced Barack Obama to become a community organizer early in his career.

It had little to do with Honeywell-style leadership. The basic idea was to find the source of what was driving people to improve their part of the world—in most cases it was some outrage or injustice in their life that they wanted to fix. The process was very agitating and difficult, especially for people like me who had no true "outrage" in their life. Throughout my life I have been blessed with a lot of support and love, beginning with my parents. Education would be my emphasis, I decided, since it had made a meaningful life possible for me, and I wanted to pass that opportunity on.

On Tuesday night of the weeklong training, 2002 election results were coming in on a lounge television. Every time a Democrat would pull ahead or win, the group cheered. My co-rider from St. Cloud asked, "Do you have to be a Democrat to be in this organization?" I answered, "No, but apparently it helps." That is one of the downsides of ISAIAH. While its positions come from the Bible's New Testament teachings, it tends to appear very political, although it never officially endorses candidates.

Fairly soon I became the chairman of the NE St. Paul chapter of ISAIAH. At the time, we had a very effective professional organizer, Doran Schrantz, who later became the Executive Director of the entire organization. Over the next few years we held town hall meetings at Galilee with the local school board and superintendent. We had two state politicians debate about voter ID and gay marriage amendments. We helped organize and attended ISAIAH's 5,000-person meetings, often at Saint Paul's large Roy Wilkens Auditorium, with over 100 state legislator candidates on the floor.

I think our most effective ISAIAH efforts occurred in 2004, when the state budget for education was finally increased by four percent after being frozen for several years. ISAIAH took the lead, with support behind-the-scenes from all the local school boards, superintendents, and teachers. I developed presentations and persuasive slides. We worked with parents' education groups and became the face of the effort, which was critical because all the others appeared to have vested interests in the outcome.

Our campaign that year was very simple. We emphasized three things: Hope, Community, and Sharing Abundance, which ran in stark contrast to the consensus emotion of the country at that time after September 11, 2001: Fear, Individualism, and Scarcity. Presenting our better values, I think, helped turn around about a 12-year history of shortchanging the school system. Most Minnesotans thought the schools were adequately funded, but for many years there had been only a one percent annual increase that let politicians say they had supported education, but which fell short of cost-of-living increases. This pattern had slowly damaged the system. There were several legislative successes around 2004, and I felt like a full-time lobbyist, visiting the state capitol over 12 times one year. Many representatives and senators called me by my first name.

There were additional ISAIAH successes in the following years on medical reform, affordable housing, domestic violence, minimum wage ordinances, and transit improvements. I was even interviewed for a book by a man named Harry Boyte.

Over time, ISAIAH became much more racially diverse, and racial equity became another important goal. The organization can still be controversial in many churches, but Galilee was a strong supporter through several pastors.

In my case, it was frustrating. I wanted to change the world, and ISAIAH was taking too long. Fixing the world is difficult, even with many people working together. Too many people still turn away from unpleasant images and inequities. I'm still searching for a magic bullet, like the image of small children being turned away from a white grade school by armed guards during the desegregation years, a picture that shifted national opinion.

A related activity for me in 2004 and 2006 became direct political electioneering. I decided to support a strong new candidate in our area, Paul Gardner. Walking the streets, handing out flyers, chatting with friendlies and arguing with others wasn't my favorite activity, but it was good exercise and probably helped. My biggest contribution was organizing his letter-writing campaign. I prepared a giant Excel table with all previous writers and their key subjects and recruited new writers. After sending out contact information and rules for "Letters to the Editor," I encouraged people to write and gave hints on what to emphasize. In 2004, during the presidential election, our efforts had a big payoff. Paul won in an overwhelmingly Republican district by only 50 votes. Two years later, he won by over 1,000 votes. Two years after that, attitudes had changed and he was kicked out of office.

The whole flurry into politics was very new to me. From my first opportunity to vote in 1972 through 2000, I had voted for every unsuccessful third party candidate who ran—an unbroken string of losers. During all my scientific and business years I considered myself apolitical. I disliked that

Honeywell always suggested the candidates that executives should vote for to help the company, and I never contributed to their election funds. I had to hold my tongue at some meetings. My more liberal turn after retirement largely came from firsthand experience with people in the poor neighborhoods around Galilee church. Learning how difficult it now is for people in such neighborhoods to break out of poverty affected my views. Strong personal teachers, encouragement, and high expectations surrounded my early years. A lot of kids don't have that and are surrounded instead by roadblocks.

In 2005, Galilee Lutheran started a neighborhood program called GLOW, Galilee Live on Wednesdays. I met the face of poverty consistently as we ministered to dozens of children from the low-cost apartments that surround the church. Every two years an entirely new group of children came to the program as families moved on to new places, often due to eviction. Some of the stories will be told later. But generally speaking, as my lobbying with ISAIAH faded, working directly with kids increased. Some of their stories were heartbreaking.

Consulting

The early years of retirement revolved around three areas: (1) Community Support, (2) Business Consulting, and (3) Writing. The consulting part could have been more lucrative, but I never truly promoted it. I built a website and let it sit. Somehow, jobs rolled in without asking.

I traveled regularly to California through the early 2000s, which was great because I could visit Amy at Caltech. My first consulting job was probably through Donna Bell, a marketing advisor who had worked with me at Honeywell. Moore Industries was that first client, a small manufacturer of devices for factories and refineries located in the San Fernando Valley. Their products had a slight overlap with Honeywell's, but I was careful not to pass important information in either direction. They asked me to review their development process for a line of new products. Employees there viewed me with suspicion at first but hanging out, sometimes for entire weeks, helped me blend in with their engineering teams.

Moore Industries was a privately held company, which meant they had different norms than Honeywell. For example, most days ended with the founder, Leonard Moore, and his top team meeting in a small glass and chrome conference room to wind down. They invited me in and everyone had a couple rounds of hard liquor to loosen the conversation. Honeywell Human Resources would be aghast at such a practice. It reminded me of the scenes from the TV show *Mad Men*, about companies in the 1960s, but without the smoking.

Moore Industries is where I met British national Jim Read, a sort of rogue engineer. Jim became a longtime friend through several interesting endeavors and adventures. A certain Thai restaurant was Jim's favorite for lunch, where he enjoyed insulting the waitress (who didn't understand English well) in the most pleasant and surprising ways. She probably considered him her favorite customer. His manner was somehow utterly blunt and completely charming.

Jim got all the side engineering jobs at Moore Industries, the new product dreams, and I was slowly drawn into those dreams. Jim later left Moore Industries, but our early joint projects were a lot of fun. One involved creating a new company called Hyperfine. Hyperfine was to be the marketing and management arm of a joint venture called TTC, The Timing Company. Jim had connected with David W. Allan, a rather famous engineer whose career had been at the National Institute of Standards and Technology. He and his colleagues developed the Allan Variance, which is at the heart of international time-keeping algorithms. Dave believed he had discovered a new way to use background variances in GPS satellite signals to make exceedingly accurate master clocks, and how to distribute that time throughout a network. This capability is critical in bank transactions, GPS for cell phones, and many other applications. I helped determine the size of the potential business, and Moore Industries invested about $1 million in the venture.

We started making trips to the mountains in Utah to work with Dave Allan at his house. Dave is an amazing character. His home, at a 6,000-foot altitude, uses virtually no power. Ten types of solar principles are used to heat the home, even during winter.

Dave is a vegetarian and devout Mormon leader. The interactions between Dave and Jim Read, a devout atheist, were fascinating. At one critical meeting with QUALCOMM, Jim was asked to pray before our product demonstration. "God," he said, "if you really exist, please make our day go smoothly." As a Lutheran with strong engineering credentials, I was considered a neutral party in this relationship.

But it didn't go well. Dave Allen kept spending money at a high rate, and at one point Jim and Hyperfine had to go without any salaries to get to the critical demonstration stage. Jim thought Dave was working like he had at NIST, on the government dole, with no concern for cost. Ultimately, the relationship got ugly, Dave Allan was unable to prove his breakthrough would work, and everything collapsed.

During all this, Jim Read and I took interesting business trips to Park City, Utah, and the Utah Olympic Park (for a conference), Oak Ridge National Laboratory in Tennessee, San Bernardino, the Mayo Clinic, San Diego, etc. New opportunities came out of those visits while we tried to sell our revolutionary "Time Bus," which never materialized. Jim Read remains a good friend, and surprisingly, after a nomadic existence as an engineer, he is now a happy rancher in the wilds of Arizona.

Another series of consulting jobs in San Diego were due to my pal and former employee, Henri Hodara. Recall that Henri managed the remnant of a company he founded called Tetra Tech. In my last years at Honeywell, I helped broker the sale of Henri's small fiber-optic group to a larger San Diego company, Ipitek, whose CEO was Michael Salour. Surprisingly, around 2003, Michael asked me to evaluate their entire line of fiber-optic sensor products. Salour and I developed a deep sense of respect during this project, and we still exchange Christmas notes. But Henri was the amazing character who created the relationship.

Henri's wife died rather early and he threw himself into his work, both at Tetra Tech and as founder and editor-in-chief of *Fiber & Integrated Optics* magazine. A brilliant scientist, he developed underwater submarine detection systems and secure fiber-optic landlines, and is apparently tireless. I attended his 80th birthday party with dozens of his friends, after which he founded SymbiOptix Corporation in 2011, and is still working and going strong at 88 years old. I only hope to last that long.

A few other consulting contracts occurred in medical technology and in radar systems. But I missed one big opportunity with a company called Yugma, which is a Sanskrit word meaning confluence. This Minneapolis software company had developed a secure way for computer screens to be linked. For example, a service technician can temporarily see and take over a computer anywhere in the world, if allowed by the user, which greatly simplifies service work. Computers can also easily work together with the Yugma software. When I visited their company, there were only two employees and I thought a large company like Microsoft would eventually overwhelm their effort. So I declined being on their Board and making a startup investment. At this point Yugma has about 200 employees, so I missed a great growth opportunity. They do have competition, from giant Cisco Systems, but seem to be holding their own.

Writing

Possibly my interest in writing began while teaching one of my teenage Sunday school classes. They were a curious, bright group; three of the boys would later become engineers. I often talked about scientific/religious themes, in stark contrast to my co-teacher who was more of a fundamentalist. He would try to explain how trees in excavations proved that evolution was incorrect and that the earth was only 3,000 years old. I gave the class counter-arguments and explained how carbon dating worked. Somehow we got along well, were friends, and the students saw all sides on important issues.

Increasingly, I worried that the students didn't know how much the world would change in their lifetimes and felt a responsibility to warn them. I became convinced that people in 2100, if we were still around then, would look back on their last 100 years and decide that how the human species itself had evolved was the most surprising change. Looking back 100 years, from 2000 to 1900, people couldn't imagine jets, computers, TV, world wars, communism and its fall, DNA, organ transplants, or dozens of other events.

From my role as "futurist" at Honeywell, I had a clearer view than most people of the environment the Sunday school kids would be living in. I promised to write a book about the future and told them that all of their names would be used for the key characters. They would get a snapshot of their lives before it happened.

So I sketched out my ideas and started writing a novel. I knew where to start and generally where the book would go. Every morning for an hour,

I walked, making notes on small papers as I traveled along a paved path through the woods and around a small lake near my home, dreaming all the way about possible futures and how to make them interesting.

At home, I would write for several hours to capture the ideas with only one goal: make it super interesting to a reader. I didn't worry too much about how it all fit together. That work came later.

It is said that to become a good writer one needs to write one million words and then throw them away, taking ten years to hone the writing craft before you become good at it. I didn't have that much time, so I set out to write a great novel on my first try.

I wrote about three completely different families encountering the technical surprises of their next ten decades and how they reacted and adapted. To make it more exciting, I threw in aliens (two types) who come to muck up human evolution. People in the book begin using more and more implants, artificial intelligence, and finally alter the human DNA, with the aliens cheering them on. Somehow a tiny black hole falls to the Earth's center, so the urgency at the end is to get off the planet in any way possible before it implodes.

I tried a conventional approach to publishing the book: finding an agent, having him or her sell it to a publishing house, marketing, and getting it into thousands of bookstores. I sent samples of the story to 20 of the 40 top science fiction literary agents, and about 60 percent wanted to see more. Sending off several chapters always seemed to be a sticking point with the agents, and one by one, they all found a reason to say no. Generally, only one in ten agents will even ask to see more of any new book, so it wasn't completely discouraging, but I think being a first-time author at my age was a major problem. Someday soon I will return to *Gardeners of the Universe*.

My writing career wasn't a quick, one-shot effort. It extended over four years, with writing classes at the local literary center (The Loft), discussions with three published science fiction authors, and reading hundreds of books. I studied the publishing process and edited my book dozens of times. Probably 20 friends read and liked the book, and I incorporated all their suggestions. Then, after awhile, I moved on to other projects and let my book sit for ten years.

Two people along the way became good friends, Phyllis Moore and

Cassandra Amesley. One of our first classes at the Loft returned rather brutal criticism to both of these women, and I later joined a small group they formed to help one another. Phyllis has since self-published two novels: *Pegasus Colony* and *Storm's Coming*. For over five or six years we met and slowly connected our lives. I was often the person who helped Cassandra put up shelves and untangle computer problems.

We usually met at a Starbucks in a Barnes & Noble bookstore, which one time resulted in destroying my car. I had a silver BMW Five Series at the time, and during our book discussion a terrible rainstorm developed. We were forced to stay in the store for an hour longer than planned. When I reached my car, I found it resting in 18 inches of water. It had been parked at a low spot in the parking lot and the drain there had plugged. My car started, but my power seat started to move forward until I was shoved against the steering wheel. I thought I was about to be crushed by a car seat, but the motion stopped just in time. The car sputtered all the way home, but I made it. The next morning it was dead—totally dead. An insurance agent said it wasn't worth trying to fix, so he soon sent me a check for $20,000, which was more than I thought the car was worth at that time. I bought another BMW and sold my spare wheels on eBay for $600. It was fun watching people's expressions as I rolled the wheels into a post office, one by one, and put them up on a scale for mailing. They just barely made the U.S. Mail weight limit.

At one of my Loft classes, I found a flyer on our chairs inviting input to a new publication to be called *The Saint Paul Almanac*. They wanted short stories that illustrated life and history in the city where I grew up. So I wrote a few pages and called it "The Dump," about the woods behind my parent's house. Surprisingly, in a competition, it was selected for publication. The following year I interviewed Miriam's mother, Helen, and submitted a story about her growing up in St. Paul. Then the following year I co-authored a story with my friend, November Paw, about her immigration to Minnesota, that was also published. By the third year, only Garrison Keilor had published as many stories in the almanac as I had.

Writing each morning is a wonderful life for a retired person. The walks are healthful, the creativity and dreaming very satisfying, and I am forced to learn new ideas on a hundred subjects for authenticity. That's partly why I started this book, to recapture that lifestyle. My top goal now is to actually get these books published somehow before I expire.

Kelsey and Joe

Our daughter Kelsey has led a rather charmed life, as if God or someone had planned the sequence. Her six years after high school were connected like a movie script.

We drove her to her first setting at Valparaiso University (Valpo) and spent several hours huddled with students in the dorm basement while a tornado passed over. Kelsey was calm, already being familiar with tornadoes. After the storm cleared, we bought her a rug, shelving, and set up her new computer while she began to make a dozen new friends, which is her specialty. One stood out, Aubri Johnson, Kelsey's "twin." At least that's what everyone said. That first day they'd both worn bright yellow shirts, blue shorts, and with matching blond hair they were destined to be roommates for their entire four years at Valpo.

Kelsey picked her major, Public Relations, and during her first semester found pals for life, and a boyfriend, Mike Rosenwinkle. These close chums traveled to Chicago just to ice skate or attend a concert, took cruises together, and later attended each other's weddings. They'd easily fly 1,000 miles on a whim if a friend were involved. Kelsey joined a public relations association and flew to a conference in New York where she heard a speech by Donald Trump on how to influence people. Persuading people was never a problem for her. Before graduating, she worked in the student center, organized Habitat for Humanity rallies (sleeping on the streets in winter in cardboard boxes), and helped build local homes. She received one of two Valpo University Service Awards.

In 2002 she was roped into being a Princess and Queen candidate for the Rice Street Festival and parade in St. Paul. After a fashion competition and long interviews she was still a strong candidate for the title, despite being totally unable to fulfill the Queen's duties. She repeatedly pointed out to the judges that she wasn't living in Minnesota anymore. Happily, her best friend, Amanda Melquist, became the Rice Street Queen, and later the prestigious Queen of Snows for the St. Paul Winter Carnival.

Valpo offered a study abroad option and Kelsey traveled to Cambridge, U.K., in January 2004 for a semester of classes there. With a new group of friends, she began to organize weekend trips throughout Europe that included Spain, Italy, Germany, Austria, the Czech Republic, Scotland, Norway, Wales, and much of England. One group of friends flew to Paris with her, where they ran a half-marathon.

Miriam and I visited Kelsey in Cambridge, and I made a special effort to get inside the Cavendish Physics Laboratory to talk to a professor there. Cambridge is a cathedral of knowledge for the entire world. If they had had Nobel Prizes in the 1500s, Cambridge would have about 200 by now. That day, our visit started at Trinity College, where Newton made his discoveries, and we saw his original papers in the famous Trinity Library. We kept an eye out for Stephen Hawking but without luck. Crossing the river Cam, we walked along a flowered arbor path and beside a pond before reaching the Cavendish Lab. It was wonderful to see the equipment used by J. J. Thompson and Rutherford to discover the electron, radioactivity, and how the atom looked. The meeting with a low temperature professor there was a disappointment, but the reservation did get the family into the hallowed building where much of 20th century physics began.

During the summer of 2004, Kelsey did public relations work for U.S. Congresswoman Betty McCollum. Then in the fall, she moved off campus into an apartment with her friends. Our entire family went to Valpo for her big graduation in May 2005. Needless to say, we were all very proud of her.

She had decided to do intern work with Habitat for Humanity right after college, so she and I drove down to Americus, Georgia (Habitat's small-town headquarters), soon after her graduation. We were a bit lost and Kelsey drove around town while I had my laptop open searching for Wi-Fi connections. There were dozens in the city center, but all were locked with passwords until we drove by a JAVA café, where we could finally pull up local maps. After I left, Kelsey stayed with several new friends in a house that had been built by Habitat and began working in the public relations department at the headquarters.

Americus, a town of about 15,000 people, is filled with Victorian and antebellum architecture and located only 10 miles from Plains, Georgia, hometown of Jimmy Carter, the 39th U.S. president. Kelsey attended Carter's

Sunday school class at his church in Plains and had her picture taken with him. Carter continues to be an active ex-president with a large foundation in Atlanta and close connections with Habitat for Humanity.

During Kelsey's internship, Habitat received a huge $100 million, four-year gift from Thrivent Financial for Lutherans (a non-profit fraternal company headquartered in Minneapolis). This thrust Kelsey into a unique position. She commented once, "I think I might be the only Lutheran in Americus." She began to play an important role in rolling out the new Thrivent program, which soon set up operations in many states to encourage Lutheran churches to help build homes with Habitat. One huge PR job for her was the Jimmy Carter Work Project, a weeklong event building about 20 homes with Carter and other famous people pounding away on boards.

All of this led to her first paying job in Indianapolis, working for Thrivent, where she tried to persuade churches across Indiana and western Ohio to help build Habitat homes. It also led to a whole new group of friends she would stay close to for years. Her PR work there included organizing a half-time program for an Indiana University basketball game and meeting with Tony Dungy (famous football coach). She also worked in Los Angeles on a nationally broadcast Thrivent Builds/Habitat promotion with famous actors, and built homes destined eventually for New Orleans to help after the Katrina floods. In L.A., she liked Mr. T the best among the actors and found some others annoyingly egotistical.

Miriam and I participated in a Kelsey-led Habitat build in a small town in central Mexico in 2006. People from Indiana and Minnesota (including our neighbor Nancy Melquist) gathered in one of my favorite towns in the whole world, Guanajuato (which I pronounce "wanna-wattle"). It's a fascinating place with all but a couple of roads underground in tunnels left from the silver mining that once thrived there. Above ground, the buildings and houses are all brightly colored in a patchwork around the white-painted university located in the center. I had a field day with a new camera throughout the trip.

We actually built new homes in a very small village (perhaps 150 people) outside the city of Villa de Santiago, where we slept. Near our hotel all the locals seemed to eye us with suspicion. We were probably the rare gringos to be seen around there. But the people in the small village loved us. Children in the schoolyard yelled and ran smiling as we passed. Everyone there had a piece of the action and had committed to help build all the new homes. They were happy to see help from Americans, and we slowly became their friends. The location was gorgeous. Herds of goats wandered along the roads, ancient wagons pulled hay through town, and a dozen volcanoes scattered around the area lent an ominous air.

Miriam and I later participated in two other weeklong Thrivent/Habitat excursions worth describing. On a Jimmy Carter Work Project in 2010, the goal was to revitalize the neighborhood where I grew up in St. Paul. In fact, we worked on a new house only one block from my boyhood home at 950 Burr Street. The whole area, and many others, had been devastated by the housing and financial crisis of 2008, and many homes had boarded-up windows and doors after being foreclosed, including my old house. The old place had been painted bright blue, and marigolds were growing where the front lawn had been. The broken-down garage had been repaired, probably all in a hope that the house would be purchased. Even the "Dump" out back had been partially filled in. But with a NO TRESPASSING sign on the house door, it was all very sad, with so many suffering through that economic collapse. I could grieve for my old house, the struggling local grade school, and the whole troubled area from the top floor of the Habitat house where I worked.

The other memorable Habitat build was near Kalispell, Montana, in 2012. But before I tell that story, I need to introduce an important new character, Joe Midthun.

As Kelsey ended her time at Valpo, she grew away from her boyfriend, Mike Rosenwinkle. While Kelsey was in Indianapolis, Amy suggested she meet Joe Midthun, one of her husband Dan's cousins. He had also recently broken up with his girlfriend, and so a series of long phone calls began between Kelsey and Joe. While running up a huge phone bill they found they had much in common: both were Lutherans, runners, energetic, and funny. Joe started flying to Indianapolis, and Kelsey saw him whenever she came to Minnesota for holiday breaks.

I gradually learned more about this new guy, how he had grown up in Papua, New Guinea (while his parents were missionaries), and attended middle and high school in Gonvick, Minnesota, a tiny town on the northern plains. Despite this weird history, I soon learned that he was a normal person and very likeable, in fact. Missing the chance, or desire, to put him through the stern father-of-the-daughter questioning, I grew to appreciate him and his whole family (after playing poker in their dining room). We weren't too surprised when Joe and Kelsey showed up with a ring on her finger in September 2006.

Kelsey moved back to Minnesota in April 2007 for a new job as a Publicity Specialist with Thrivent in Minneapolis. Joe and Kelsey started looking for their future apartment and getting ready for their wedding, scheduled for October. Joe had enrolled at Luther Seminary after graduating from St. Olaf University and, like many others in his family, began studies to become a pastor.

The actual wedding was on October 6, a glorious fall afternoon, at our Galilee church. The deep red dresses of the bridesmaids complemented the autumn leaves that surrounded the church grounds. A reception at the "glass palace" Como Park pavilion, with a small penguin as a special guest, celebrated the event with dancing late into the night. After a short honeymoon in Cabo San Lucas, Joe and Kelsey began authentic married life.

The next year, 2008, was frenetic for both of them. Joe started an internship ministering to athletes and other students at Augsburg College, while Kelsey strengthened her connections to Habitat for Humanity with trips to Madagascar, Nicaragua, El Salvador, and New Orleans. Looming in the background was the question of where Joe would end up as a pastor. A story Joe tells often is how he interviewed at All Saints and Ostmark Lutheran, two churches in west-central Minnesota. They had struggled with their pastors for many years and were discouraged by their recruitment results. Joe was the last of six people to go through the interview process. But they must have seen something special in Joe because, on the long drive back to the Twin Cities, he got the phone call asking him to be the new pastor for both churches.

Interest rates and home prices were at historic lows when Kelsey and Joe searched for a new home close to the two churches. They found a great house overlooking Spring Lake near Dassel, Minnesota, with a beautiful view of the evening sunsets. Kelsey continued at Thrivent in Minneapolis, carpooling for over an hour twice a day.

Now that you know Joe, I can tell the story about a nifty weeklong Habitat build in which Miriam and I participated. Joe and Kelsey led a team with people mostly from Joe's new churches. We slept at a Lutheran Bible camp on Flathead Lake, near Kalispell, Montana, and many evenings we sat on a porch facing the ebbing campfires on the lake's edge as the sun set. Joe spoke those nights with a new serious voice, like a pastor, explaining his view of the Bible and humanity. He seemed quite different from the funny, concise son-in-law I had gotten to know. Joe betrayed his depth and seriousness to me on that porch for the first time.

We worked on two new homes in the relatively flat region just west of the Glacier Mountains. Some mornings we watched the fog lazily spool over the mountains in the orange sunlight as we prepared for work. Safety is a high priority at every Habitat build and a precautionary lecture always precedes the work, so it was surprising and unfortunate that a parishioner from Joe's church was accidentally shot with a nail gun. The nail in her thigh had to be dug out. It must have been a clean nail since she healed well.

Miriam mostly built and painted walls while I helped construct forms for the foundation pour. It took three days to build those wooden forms with metal pins to allow them to be easily removed after the cement cured. On the last workday, when the cement trucks finally dumped their loads, Joe walked along the top of the forms pushing the wet cement down, removing bubbles, and showing his dexterity, balance, and courage on the high wooden structures.

We met the families who would move into the houses during a lunch break and heard their moving life stories. Feeling good about our accomplishment, we had the final day off and drove to Logan Pass in Glacier National Park on the "Going to the Sun" road and hiked up into the ice fields. Kelsey, Joe, and several others slogged even further over the ice crest, while Miriam and I stayed below and learned to slide down on the ice while standing.

Early in our marriage, Miriam and I had talked about biking through the Canadian Rockies, probably the best ride in North America, but we had never followed through. Now too old to bike up mountains, we were determined to do it in a car. So when the Habitat build team flew or drove home, we stayed on for a week's vacation in the U.S. and Canadian Glacier Mountains.

My new camera got a workout as we again crossed Logan Pass and stayed near beautiful St. Mary Lake. Driving north into Canada, our first stop was at a large, picturesque facility known as the Many Glacier Hotel, where a series of ferry rides and portages led to a very high glacial lake. The first ferry ride was on Swiftcurrent Lake, which emptied its waters at a waterfall under a bridge near the hotel. About a hundred yards out on the lake the ferry's engine failed and we started drifting toward the waterfall. A small boat was sent out to rescue us by starting the ferry engine, but they were unsuccessful. We kept drifting, getting close to the drop-off. Finally, the boat staff gave up trying to start the engine and got another boat out to pull us back to shore. Miriam and I were in a small group who decided to wait around on the dock for an hour to see if the engine could be fixed, and we were royally rewarded. The ride and hike up to the high lake was gorgeous.

It's difficult to describe how beautiful the rest of our ride north into Canada became. Our itinerary included: (1) high tea at the window wall in the Prince of Wales Hotel, overlooking another glacial lake surrounded by mountains; (2) a walk along a famous peaceful path beside Lake Louise; and (3) a drive toward Jasper with an excursion to the Columbia Ice Fields. If you ever have the opportunity to make this journey, don't put it off.

Life and Death

Amy announced her first pregnancy in June 2003. We were all nervous when Amy said she planned to have a home-birth with midwives only. Many older women tried to talk her out of it, citing the dangers of not being in a hospital if something went wrong, but Amy was resolute. She felt that hospitals were too quick to use special procedures to reduce their risk, especially insurance risk. As a result, she explained, too many babies were born using caesarean section or with various drug treatments. She wanted a more calm and natural birth in her bed, and of course, she did all the research.

Peter Kenneth was born around 7 a.m. on March 8, 2004, after about four or five days of labor. Amy had gone on walks and tried every trick, but long labors may be a family trait. Her own birth had taken two days for Miriam. When Peter finally arrived, he experienced seizures and breathing problems and was rushed in Dan's car to the Neonatal Intensive Care Unit at Children's Hospital. Similar problems occurred two years later, on

March 12, 2006, when Leo William was born, at which point Dan said he couldn't take the stress and terror of a dangerous delivery anymore, and the next birth, Tim's, began in a hospital.

Peter and Leo were beautiful babies and toddlers with amazing smiles and personalities—and I'm trying to not be too biased as their grandfather. They were both mobile very early and risk-taking climbers (especially Leo, who eventually visited hospitals himself a few times after falls). I like to think the climbing and jumping came from Ronald genes. Like Amy and Kelsey before them, they were perched on our refrigerator top at only a few months old to topple and fall into my waiting arms.

My mother Fern loved Peter, her first great grandchild. The entire family vacationed at Rutgers resort in Deerwood, Minnesota, during Peter's first summer, and Fern would hold Peter and they would laugh at one another, staring at each other's face. She was 91 years old that summer and beginning to slow down a little. Toward fall, I had more difficulty getting her out for walks until, finally, she just didn't try any longer. Her last birthday, in November, and final Christmas were happy, but in the new year, 2005, she seemed to only want to sit in her favorite lounge chair with a blanket over her.

At Summerhouse in Shoreview, Janis, Miriam, our daughters, and I were able to visit with her more often. On January 9, I spent the evening with her, making supper (a small pizza, salad, and fruit), and we looked at photos from her youth up to the latest ones. At midnight, after I left, she phoned and asked how to turn off her TV (she had never fully mastered the remote), so I told her. The next morning her friend Doris called and said she thought Fern had died in the night. We rushed over. I found Mom unmoving, still in her favorite chair with her mouth slightly open. The police and doctors said she had died of natural causes.

For my father, my first great loss, the black hole of his absence lasted for at least ten years. I suppose it was a little less devastating when Mom died. She had lived a long life and her death was somewhat expected.

Mom had devoted herself to her family, especially Janis and me and our education. At Summerhouse, she refused to give up her car, despite the $30-per-month garage fee because, she said, she might need to drive over to take care of me if I became ill. She hadn't driven in at least five years.

She expected little of me except to love God and stay in Minnesota. She had seen the world transformed. Born before World War I, she lived through the Depression, World War II, jets, computers, TVs, and men standing on the moon. She loved me, and I loved her completely.

Decade of Chaos and Consequence

The time period from 2000 to 2010 is when, in my opinion, chaos became the norm. The late 1960s had included tumultuous events, but by the 1990s, things had settled down to "normalcy." Sure, there were small wars and serious fights in congress, but the Berlin Wall had fallen, the economy steadily increased, and people generally had jobs.

Many prepared for a catastrophe that never happened at midnight, December 31, 1999, when the new millennium began and all the computers were expected to go berserk due to clock errors. I didn't worry much, and the computers rolled over without a disaster.

Then the real troubles began. The dot-com bubble bursting in March 2000 was merely a pre-shock for something much bigger to come on September 11, 2001. The World Trade Center and Pentagon terrorist attacks shook the country's sense of security. Soon after, the U.S. went to war, first in Afghanistan (probably justified) and then in Iraq (crazy stupid). The U.S. military's quick march to Baghdad seemed satisfying at first, but no one had worried about what to do with that broken country afterwards. Soon fear became the way to win elections—the fear of Islamic terrorists, the fear of walking at night, the fear of losing your job, the fear of the future.

I experienced shocks as well. Around 2004, my long-term doctor sent me to the University of Minnesota Research Hospital to see if anything could be done about the blockage in my portal vein (which was still forcing me to have esophagus treatments every six months). Dr. Butler also was worried about the mini-stroke I experienced during the esophagus procedure in 2002. The university doctors said there were 14 new possibilities, and after long discussions we decided to kill off part of my spleen. Most of the blood that tries to flow into the liver comes from the spleen, which was causing many of my problems. So they counseled me to have a "simple procedure" in which parts of the spleen would die, reducing the blood flow. (Another option considered was simply removing the spleen, but that damages one's immune response.) The simple procedure involved inserting tiny plastic particles through a tube from my leg into the spleen, shutting off blood flow, effectively killing part of the spleen. The official name is partial splenic embolization. I would need three treatments to make the spleen as small as desired. I said okay, having no idea what it all meant.

Each treatment meant that part of the spleen was really dying, rotting away, just below my diaphragm. Every time I took a deep breath, the croaking, tender, raw spleen would scream out at me. After each treatment, I could not lie down, so for six weeks I slept sitting up on the living room couch. Needless to say, my lifestyle was pretty upset by this yearlong pain, but ultimately, the treatment worked. I went without an esophagus stretch for the next 10 years. The whole episode was a reminder of the fragility of life, which caused me to take on a slew of new projects and to work twice as fast once I'd recovered. For example, that's when I decided someone needed to organize and digitize our 80,000 family photos, which will be described later. I began thinking about death more often.

Other shocks in the decade included Miriam finally retiring in 2005, which added several new home, church, and family projects to my list. Generally, I was moving away from organizations and being organized, and just started doing whatever need was currently in sight. It turned out that a lot of people needed my help.

Technology shocks also occurred. In 2007, the iPhone and Android smart phones appeared, Facebook and Twitter went global, and the cost of analyzing DNA and genomes shriveled. Changes in communications, in particular, have altered how humans interact, organize, and their power structures. In a sense, they have changed humanity itself. The new social media inventions of 2007 made it possible for one person to dominate the attention of the world—for example, a Donald Trump, a vicious terrorist, or a democratic revolutionary. One only needed to ignite an issue or a cause, go viral, and capture enough followers.

Before the decade ended, the country almost entered another depression from the housing/insurance/Wall-Street implosion. We helped some friends through that crisis with direct financial help. I taught people from the church neighborhood how to search and apply for jobs on computers. I saw firsthand how the lives of poor Americans were blocked on all sides: in housing, transportation, food, jobs, and most importantly, education. They were also falling further behind due to the expense of the technology revolution. The expectation that they would pull themselves up by their bootstraps seemed like a joke in most cases. The last half of the decade for me became a messy tornado of reaction to the chaos and helping others.

My politics solidified at this time. Seeing the desperation of poor people of all colors was part of the reason. But the impulse to help others was only part of the picture for me. People needed jobs and self-growth. I knew, from my years as a business leader, that companies don't hire when they have a good financial stretch, because the next quarter the stock market will expect another 15 percent growth over that previous period. So unless a company has a sure-thing product ready to go, they will use their extra cash to buy new technology to reduce costs rather than hire more workers, or they will save it for the next quarter. And, as a relatively rich person, I know that people who are well off don't spend all their extra cash each year. They save it. So the economy, in general, doesn't grow. What does cause hiring is when the demand for products is so great that a company "has to hire" to just keep up. And the demand for product is greatest when poor people have extra cash, because they spend almost all of it. So, in my opinion, the economic system in the U.S. is broken, and some realignment of the tax system is essential to get it growing again. And this doesn't mean a tax cut for the rich.

Out of the chaos of the 2000s came a hundred projects for me and tugs in all directions. The stories that follow are those that rise out of the muddle.

The Karen

I need to tell you the story of the Karen, that's pronounced Ka-rén with the accent on the second syllable. The Karen once lived mostly in Burma, now officially Myanmar. About 20 percent are Christian, with the remainder mostly Buddhist. Generations of British and Baptists helped educate the Karen and developed their sense of identity, before a 60-year civil war started around 1950. The Burmese military government began to force them out, chasing them from subsistence farms in the eastern part of the country.

Many fled across the river border into Thailand where they live in several large refugee camps of over 50,000 people.

The Karen entered my life around 2005 when I got a call from Jill Lund, a member of the Roseville, Minnesota, school board. We had met at one of the ISAIAH town hall meetings to restore school funding around 2004. Jill and a woman from the Roseville schools, Peg Kennedy, described a mini-crisis. About 150 new children, all Karen, had shown up at the bus stops, expecting to go to Roseville schools that fall. The school system had no warning and the Karen families had no idea what a Minnesota winter would be like. We decided to invite all the local churches, about 150, to Jill's large church building for a presentation by Peg about the situation, hoping that each church would "adopt" a family and help them adjust to Minnesota. I had the addresses and pastors' names for the local churches and agreed to invite them all. At the end of a big evening, 35 churches and individuals had agreed to serve as sponsors and tutors for the families, including Miriam and me.

That's how we met the Paw/Tunhla family. Two of the Paw daughters, November and December, knew just enough English so we could meet with them without a translator. Soon, I had given the family one of my used computers and installed a Rosetta Stone application for learning English. The children got right at it, and within six months were speaking passable English (the ESL classes at school probably also helped). November was soon mainstreamed at Roseville Area High School, and she ultimately graduated in the top third of her class. A story we wrote together for the *St. Paul Almanac* told of her life in the refugee camps, how her father had been shot through his hand escaping from Burma, and her harrowing introduction to America, with two days of starving and sickness on the trip from their refugee camp to Minnesota. November read her published story in coffee houses. We helped her get a Pell Grant to study at Century College. She later earned an Associate of Arts degree there.

Every member of their family is amazingly nice and we became true friends over the years. They came to our house and played with our grandkids. We helped them through a bedbug infestation. Warm, happy, and hardworking, November never forgets to thank God for their lives. But her proudest moment, with December, was when they became official American citizens in 2013. It's hard to describe the joy in their eyes as they said the Pledge of Allegiance and later posed with their American friends. It's a moment I can't forget.

I often think about my own family's assimilation into America, which took about 70 years in my father's case. Throughout his life, his English was strongly accented, and not until his children's college years were we fully competent grammatically.

In our time with November's family, our goal was to speed up that process. As with all immigrants, some children will succeed and become educated, some will work at humble jobs, and maybe some will join gangs and have bad lives. But these families bring new vigor and life to our country and, given some time, become the artists, entrepreneurs, and leaders of our future. The Paw/Tunhla family has now purchased their first house. Knowing them has been an enlightening and valuable chapter in our lives.

The Most Amazing Day in My Life

My activities with church and community peaked around 2008 when I went to a Lutheran St. Paul Synod meeting and was somehow elected to represent the North Central region at the national Lutheran convention, officially called the Churchwide Assembly. As such, I was representing about 30 Lutheran churches in the northern part of St. Paul and its suburbs. Over a year passed between this "election" (actually more like whoever volunteered) and the real Churchwide Assembly, which was scheduled to take place at the Minneapolis Convention Center in August 2009. Over a thousand delegates would represent the 11,000 congregations of the Evangelical Lutheran Church of America (ELCA) and their 4.7 million members.

I spent time throughout the year leading up to the assembly reading, talking, and listening about what was sure to be the defining issue up for a vote, namely a new church position statement called "Human Sexuality: Gift and Trust." This position statement took up the issue about whether ordained people in a committed same-sex relationship could serve as ELCA pastors. At the time, and perhaps still, it was a highly controversial issue. The official statement admitted that individuals and churches had what they called a "bound conscience" on the question, and that people disagreed for very serious reasons that should be honored. Churches would be free to decide on their own how they would answer the question, but the national organization would not impose a hard rule. The position meant that for the first time some congregations could choose to have a gay pastor.

Miriam served as an attendant at the sign-in desks. Admittance to the main voting area was strictly controlled. For several days, routine bureaucratic matters were decided, but everyone knew the big issue was drawing closer. I was staying in the nearby Hilton Hotel, sharing a room with a pastor from Ohio. Every night we debated, sometimes for hours, before falling asleep. In a nutshell, his position was that homosexuality was sinful and that no pastor should be allowed to repeatedly commit a sin. For example, he said that if he swore, taking God's name in vain during each of his sermons, he would surely be tossed out of his job. I felt that being gay was a condition in some lives, perhaps caused by a genetic or hormonal glitch during pregnancy, i.e., that most gays are *born* gay. There does seem to be a persistent ratio of gays to straight people of about eight percent over eons. My roommate eventually agreed that that made a lot of difference in how to think about the situation. We went to the big voting day, however, still on opposite sides. Neither of us envisaged the astonishing events that were about to occur.

On the fateful day, I was seated in the front row among all the delegates about 25 feet from the presiding bishop, Mark Hanson. We gained the front rows because St. Paul and Minneapolis were the host synods at this General Assembly. The new position statement on sexuality and its implementation procedures were debated for hours, with person after person speaking, including a former governor of Minnesota, Al Quie. I sat, attentive, listening to the arguments. At last, the vote was to be taken. The doors were sealed. Anyone visiting a restroom would be left out. Throughout the Assembly we had been voting using hand-held electronic devices, which had often been a problem. People had complained that their votes had been missed because not enough time had been allowed. To pass, this new position statement would require a two-thirds vote of all the delegates. So when we finally pushed our buttons, Bishop Hanson looked frightened. He fidgeted for a while and said into the microphone that he needed to talk to his parliamentarian. The conversation went on for at least five minutes before the Bishop again moved to the microphone. He admitted that he had been delaying so that no one could claim that his or her vote had been missed. Then he announced that precisely two-thirds of the delegates had voted yes. If a single vote had been changed, the result could have changed. In a sense, every person voting yes, like myself, was responsible for the result.

The vote was announced around mid-afternoon, and people were still recovering, and perhaps complaining, when an announcement was heard throughout the building: "Move away from the hallway windows, a tornado is approaching." Miriam and all the other attendants moved into the delegate room with us while a terrible roar and rush passed by. We later learned and saw that the domed roof on a different section of the convention center had been damaged, and across the street, the steeple of Central Lutheran Church had been ripped off. We were all happy to escape the afternoon unharmed, but I suspect most of the people present speculated on the extreme coincidence that had occurred. Many left convinced that God had spoken, and ultimately about 15 percent of the Lutheran churches at the Assembly decided to split off and form their own new group of churches where a ban against gay pastors stayed in effect. But overall, the Evangelical Lutheran Church in America recovered and moved on. Like all mainline churches, the ELCA is slowly losing members as younger people move away from religious institutions in general. But I am proud of my vote and the stance of the ELCA in accepting all varieties, shapes, and rascals into a forgiving, loving family.

What I Believe

By 2010, my life had become a swirl of community action, politics, and church-related activities. For example, while I was Galilee Council President, I helped select two new pastors, Ralph Baumgartner and, later, Dana Nelson.

Throughout much of my life I'd been quiet on topics like politics and religion—at the dinner table, work, and even back at school I'd held my tongue. I was steadfastly unopinionated. It may have been an outgrowth of my natural introversion but, by 2010, at the age of 65, I'd found my voice. Perhaps I'd become old enough to know what I really believed.

How to best support the community was an area that changed for me. Improving the world through ISAIAH seemed too slow. We hosted ISAIAH discussion parties at our house, but I'm not sure how much was accomplished. In my last major activity with them I helped organize a debate among gubernatorial candidates, which took months of preparation, speaking to the campaign people, defining the format, and inviting organizations and churches. Hundreds of people attended the event at Our Saviors Lutheran Church, including the Independent candidate, Tom Horner (whom I got to know well); the prime Democrat-Farmer-Labor (DFL) candidates, Mark Dayton, John Marty, and representatives for R. T. Rybak and Margaret Anderson Kelliher. Fringe Republican candidates also attended. The meeting was a big success with students and other local citizens posing questions to the candidates. The entire event was filmed and later posted on the web. The only bad result was the total exhaustion of the members in the NE St. Paul Caucus of ISAIAH. A few months later our professional organizer, Dai Thao, left (he later became a St. Paul City Council member) and Jeanne Ayers (my co-leader) joined the new governor's Health Department. I took what became a permanent leave of absence from ISAIAH, mostly due to its intense time commitment. We really weren't changing minds fast enough to make a difference. I wanted to see results!

That spring I became an alternate to the state DFL convention in Duluth, which was pretty interesting. I sat in the upper deck watching the official delegates down on the convention floor. After several days, Margaret Anderson Kelliher received the endorsement of the convention. Mark Dayton ignored the convention results and won the fall Democratic primary. He later won the general election versus Republican Tom Emmer and Independent Tom Horner. I think his two subsequent stints as Minnesota governor have been quite successful, with excellent economic performance and rebuilding of the state budget from deficit to surplus.

My views on the economic side of politics had solidified. As mentioned before, pouring money back to the richest citizens doesn't seem to pay off. They tend to hoard their money, investing some, but also buying extravagant unproductive assets. By contrast, poorer people tend to spend most of their cash on necessities, passing the money to vendors who also spend on necessities, in a positive, virtuous cycle. I think the relative success of Dayton's Minnesota and other wealthy Democratic states proves the point.

Everything, of course, is a balance. It's not wise to spend on unproductive government projects. But I think tax breaks for the rich are foolish. We currently have far too much concentration of wealth in the U.S.

Surprisingly, the activity that forced me to reevaluate my religious beliefs most acutely was a small group of teenagers in Galilee's GLOW program (Galilee Live on Wednesdays) that I mentioned earlier. The group was amazingly diverse, with blacks from the neighborhood, atheists, devout Baptists, Hispanics, gays, Caucasians, tall people, short people, kids who had been at Galilee for years and new people who were invited each week. Almost all of the 5–12 people who came on Wednesday nights were very poor and had myriad family problems. The concept of sharing was totally foreign to all of them. Incidentally, they also all believed in ghosts.

About this time, I was writing a blog on the gospel of Luke and the Acts of the Apostles. One of the concepts I explored was (1) exactly who was our "neighbor" and (2) how much were we willing to do for our neighbor. I pointed out that for Jesus the people of the whole world were his "neighbor" and that he gave his life (everything) for others. I made a big chart and invited the GLOW students to put a dot or line where they stood on sharing with others. Beyond their families, they were unwilling to do much of anything. As I slowly learned the stories of their lives, the poverty and struggles were shocking to me.

One girl, the atheist, illustrates the situation. She was amazingly smart and cynical about religion. I learned later why she felt so strongly, after visiting her mother, nominally to help them move. They were being kicked out of their second apartment because the grandma who lived with them had a lung disease that required an oxygen tank, but she refused to stop smoking. The landlord didn't like that flammable combination. So when I arrived to help, the mother said the police had arrested my student for punching her (the mother) in the face. The mother blamed all their problems on the teenage girl because she didn't believe in God. "God is punishing us," she said. Clearly, the mother had bullied her daughter about this for some time. It was easy to see why the girl was such a skeptic.

So how do you talk to a class with wildly different ideas about God and religion? The key question for them was "How could God allow so much suffering in the world, in 'my' world?"

Here is what I told them. God, if he exists, wants people to have free choice. He has designed a universe where it is virtually impossible to know for sure if he exists. If it were an obvious answer, there would be no free choice, so people have to simply decide if they are believers or not. There are a hundred "why?" and "why-not?" reasons to believe in God, some of which I've listed in the following table. My growing up in a mildly religious family, where the issue was never in question, and then going to perhaps the foremost scientific and God-skeptical university in the country, I decided that it is impossible to know with scientific certainty. That probably makes me an agnostic, but I have simply *chosen* to believe and have faith that God exists. That was my decision, but I have no way to definitively prove it to anyone. And that's what I told the students, that they have to decide for

REASONS TO BELIEVE IN GOD	REASONS TO NOT BELIEVE IN GOD
Because my parents and 2,000 years of history have supported the idea.	All those people could be wrong. Some memes never go away.
Because of the New Testament and the verifiable details about Jesus.	We know Jesus existed, but beyond the Bible, do we really know that he rose from the dead?
Because of the beauty, poetry, and profundity of Bible stories.	People had centuries to edit the book and reinforce their beliefs.
The radical perfection of Jesus's life and teachings.	It's not necessary to be religious to believe in love and helping others.
The beauty of nature, nature's rules, and the universe.	The universe is an empty, gigantic place where we are a tiny insignificant part.
Due to the impossible perfection of human life.	Evolution tends to drive improvements over very long periods of time.
Because of music. Only God could make anything so beautiful.	Humanity searches for beauty to make life bearable.
Because of the promise of eternal life. I want that.	We can only hope there is an afterlife. Who knows?
Because of the unknown whys: the Big Bang, mathematics, quantum mechanics, and the improbability of our universe.	The possibility of the multiverse, where we happen to be in one that works. As Einstein said, "God is Nature and Nature is God, no more."
Because it feels like there is more beyond ourselves.	The human mind has evolved to imagine and project beyond our physical vision.

themselves, that God wants each of us to decide. I told them that I simply had decided to believe.

Grandpa Camp

In the spring of 2013 I had a crazy idea. I would invite grandsons Peter and Leo to stay about a week each at our house, sort of a camp, and I could teach them advanced math and science. By this time they had both demonstrated precocious abilities and I wanted to put some structure to it all. Peter, at age five, had awakened his parents in bed one night to talk about a way to convert a fission reactor into a fusion one. Dad Dan told him to go back to bed and they would discuss it in the morning. Similarly, Leo once had awakened Amy and Dan to tell them he had designed a new atom, the Leonium, and that he had a picture of the orbital structure. It had an atomic number of 120 and, he asserted, "It's a little unstable, like me." He was also told to go back to bed.

The idea of understanding how things really work, down to the atomic level, always seemed very valuable to me. Not knowing how or why a TV, smart phone, or a million other things work seems horrible. Considering such devices as magic black boxes is just not right for modern humans. Also, I love Peter and Leo and all the other grandchildren and wanted an excuse to spend time with them. Later, I discovered my Grandpa Camp idea would also force me to learn or relearn technical skills. I still love learning and Grandpa Camp gave me a reason to study again. The camps were also known at the beginning as "Nerd Camp" or "Peterson Summer Institute." Kelsey had T-shirts made each summer to make the camp seem more official.

Grandpa Camp will be in its sixth year, and I am running out of easy topics for the older boys, so they will be asked to do a three-day super project this coming summer. Each year, I've picked subjects that were an important part of my life, and there were many. In addition to my favorite science subjects, I've taken painting classes; learned the basics of photography, movie making, woodworking, and astronomy; studied finance and stock markets; and taken a whole slew of management and human interaction classes at Honeywell. My friends say I have an unusual skill set. I also wanted to learn new areas like microbiology, modern software, and cooking. After a couple of years it became clear that two to three months of reading and preparation were needed to develop lesson plans for each Grandpa Camp. They would be learning entire semesters of subject matter in one day, so the methods and material had to be carefully chosen and presented in an exciting and compelling way.

The table on the following page shows the materials chosen, including the break at the fourth year where we simply took the boys to London. As the years passed, I increasingly dragged Miriam into the week, as well as outside experts. The general plan was intense learning for about two hours in the morning, often on large sheets of paper, with physical demonstrations, and fun activities. The afternoons were more lessons, cooking competitions, field trips, and building items to show parents. Evenings involved free time, watching instructive movies on TV, and reading cool books before bed. Computer tools were emphasized, both because the kids like using the computer and they aren't allowed much screen time at home. For example, Peter and Leo demanded that I teach them software programing the second year.

Peter and Leo actually seem to remember much of what we covered. Peter would stick the large sheets with math concepts on his bedroom walls for months. Once at a school program called "Grand-friends," at which I volunteer, I asked Leo's fifth grade classmate Landon about the number i, and Leo interrupted to say, "It's the imaginary constant, of course, the square root of minus 1."

My problem is that the younger grandchildren—Tim (age seven), Elaina (four), Linus (three), and Heidi (one) already want their Grandpa Camp time. Luckily, I have lesson plans developed from the earlier camps. I hope I have the energy to teach them all while coming up with new ideas for Leo and Peter. Maybe I will have to learn how to fish.

I'd like to introduce you to the grandkids, each unique and amazing in many ways:

Peter, now 14, is serious, quiet, intelligent, and kind. He has recently been accepted in the UMTYMP program at the University of Minnesota, a program for talented math students which accelerates their math learning. He should finish three years of college calculus by the end of his high school years. He is also a spelling bee champ and builds robots and electronic games. He recently was invited to the state history competition with his movie project on the U.S./Soviet joint space efforts.

Leo, now 12, loves to climb and take risks. He was injured three or four times before age five. (I hope he doesn't become a mountain climber.) While at summer camp last year, he was thrown from a horse that had been stung by a bee. He suffered a slight concussion and had to restrict his activities for a while. A couple weeks later he fell out of a tree and broke his wrist. Not bad for only one month. Leo learns independently and likes to be challenged. In many ways he is brilliant. He composes music, writes software, is self-sufficient, and impossible to fully control. This summer he hopes to launch a weather balloon with a carbon dioxide detector on board.

Seven-year-old Timothy is joyous and studious. I love to be with him, and last summer I tried to teach him trigonometry, even though he hadn't officially learned to multiply or divide yet. He goes to a school that emphasizes science, technology, engineering, and math, and is doing very well. He is becoming a chess expert, rides his bike at breakneck speeds, and enjoys going to Dairy Queen with me.

Linus, Amy's youngest son, at three years old is perhaps the cutest and friendliest person I've ever known (although all of Amy's kids were cute at that age). He wants to be a fireman and seems fixated on fire trucks at this point. He is growing up in a very competitive family and knows when to shout to be heard. But they all love one another and it shows.

Elaina is Kelsey's oldest girl, at age four, and she completed her first extended Grandpa Camp last summer. We studied puzzles, cooking, painting with acrylics, color mixing, chemical reactions, games, and had fun at the park, in the sandbox, and riding to Dairy Queen. She loves a book on anatomy. Perhaps we'll do some dissection at Grandpa Camp this summer. Elaina knows how to imagine and helps with household tasks. I'm convinced she has a high dose of Miriam genes.

Finally, Heidi, at age one has a natural smile and is now waddling at a fast pace. I bought a book called *Quantum Mechanics for Babies* recently, so I'm ready for her. Where her life goes is still a complete mystery, but with Joe and Kelsey as parents, she is sure to be wonderful.

So maybe I am overly proud of these new young people, but I am part of them, and they are part of me. I just want you to know how much I love them.

GRANDPA CAMP CLASSES

2013 8 YEARS*	2014 9 YEARS*	2015 10 YEARS*	2016 11 YEARS*	2017 12 YEARS*
Art Day Frames, gesso, beauty, acrylic painting, Sketch-Up architectural drawing.	**Anatomy and Physiology Day** College homework, dissecting eye, brain, heart, suturing, brain function, and neural computing.	**Bio Tech Day** Modern microbiology, methods, petri dish experiments, bio movies, blood type test.	Trip to London:	**Art Day** Photoshop training, drone-photography, paint a self-portrait, pottery lessons at Tim & Mary Fair's house.
Math Day Shapes, n-space, Euclidian geometry, polynomials, exponentials, Excel, pie charts, areas, slopes, velocity, and acceleration.	**Math Day** Probabilities, permutations & combinations, poker & dice, fractions, number types, simple algebra, imaginary numbers, sequences, vectors.	**Building Day** Wood shop training, design 3D object and submit to 3D print shop.	Lego land, Windsor, the Tube, Trafalgar, Transport Museum	**Music Day** Music appreciation, composition, Garage Band, electronic keyboard, music history lessons at Jo Ann French's house.
Finance Day Earning, buying, budgeting, compounding, investments, poker, visit finance office, business management, Monopoly game.	**Software Day** Basics of programing methods, learn HTML, design & build personal website, machine learning.	**Culinary Day** Cooking methods and science, balance, *Chopped* competition with Miriam, molecular cuisine, practice big dinner cooking.	Tower of London & Bridge, London Eye, HMS *Belfast*	**Physics & Math Day** Imaginary exponents, QM, wave functions, Hamiltonian, & Schrodinger, fine dining etiquette

Physics Day	**Physics Day**	**Math/Physics Day**		**Advocacy Day**
Light waves, spectra, speed, Snell's Law, optics, focal length, light bench, Bohr atom, build 6-inch telescope, microscope & camera optics, CCDs, photo enhancement, Ron's big telescope.	Atomic theory, conservation laws, Newton, low friction experiments, momentum & energy, satellite design, cosmology theories, constellation game, Eagle Lake Observatory visit.	Algebra review, calculus basics, numerical methods, Maxwell's equations and other key physics laws, special relativity, watch *Interstellar* movie and discuss.	Boat to Greenwich, Science & Natural Hist. Museums	Issues, capitol meeting & tour, Kelsey's workplace, government overview, macro/micro economics, lobster cooking.
Chemistry Day	**Basic Human Interaction Day**	**Electronics Day**		
Atomic structure, periodic table, chemical bonds, reaction equations, experiments, the Mole, introduction to organics, blowing up sodium.	Myers Briggs test, individual & group performance, people types, leadership & management, negotiating, writing skills, tense, point-of-view, critical reading.	Learn modern electronics elements, micro transistor tech, quantum computing, Moore's Law, build radio, visit U of M Nano-electronics lab.	Train to Cambridge, British & War Museum	PSAT test. Return to Tim & Mary's to finish pots. Finish portrait.
Movie Day	**Travelogue Movie Day**	**Presentation Day**		**Chemistry, Biology & Presentation Day**
Video camera training, Final Cut movie basics, special effects, green screen, old movies, and story board/script. Show movie to parents.	City tour to make video, Final Cut Pro training. Show to parents.	Learn PowerPoint, Excel in depth, graphics, movies & other inserts. Three-year strategic plan including college finance plan.	Kensington Palace, Parliament Square	Ga and H chemistry, dissection, present music to parents & cook supper.

*Average age of Peter and Leo

The Garden

I have always hoped that someday we could return to the garden, a perfect world like Eden, either by God's hand or through eons of humanity steadily making the world better. I have tried to be a part of that tangible human progress. But the goal is probably impossible. Due to Adam and Eve's free choice, we were kicked out of Eden, and also due to free choice we may always be at one another's throats, unable to build the perfect world. But we should try.

In one sense, I started developing the Rice Street Gardens back in 2014. A phone call from Ken Kawai, an exchange engineer from Japan I knew at Honeywell's System and Research Center in the 1980s, finally got through to me after several tries. Ken had been a young, 20-year-old guide in Japan for both my trips there. He asked to meet and I invited him to my house. I was surprised to learn that his title was president of Azbil (current name of Yamataki Honeywell) and that he was about 50 years old. He was looking for someone to act as co-leader for a new Azbil research startup in Santa Clara, California. For some reason, I said I would help find the right person. Briefly, this is how it went: I posted the job on LinkedIn, got 120 reasonable candidates, called 20, and finally had five in for interviews at the Silicon Valley facility. Our top candidate wanted too much money, but we soon hired our next choice, an excellent leader. I'm not sure being a headhunter is my favorite job, but it worked out for Azbil, and I later received great compensation for a couple months of part-time work, about $20,000 after taxes.

That's when I made a deal with Miriam. She agreed that the money I made from actual work could be used any way I liked. And my plan was to give it all away. So I started looking for worthwhile projects that needed $20,000. The first candidate was a new brick storage shed to be built adjacent to Galilee Church, but the church council ultimately shot that down. The next attempt was to enlarge the windows at the church, but that turned out to be very expensive and of marginal value.

Then, unexpectedly, Katheryn Schneider (an interesting, talkative woman) met with Galilee's pastor, Dana Nelson, and me. She said the St. Paul Area Water Board had purchased a large parcel of land across the street from our church and it might be possible to develop a community garden there. I had tried a few times to have Galilee build a community garden on our church property, but the idea had always been rejected for good reasons.

Soon a long sequence of struggles and miracles occurred, which I will summarize here:

- The management at the Water Department liked the idea and wrote a lease agreement for us at no cost.
- A council of local groups formed that supported the garden idea and helped with details.
- We gave five presentations and passed expensive construction permits for the City of Maplewood (where the garden would be located).
- We got permission from an adjacent pub to use their water pump if they could be connected to city water (which eventually happened).
- Several hundred local immigrants and apartment dwellers showed interest and learned our rules.

Throughout this process I led many of the technical, engineering, financial, and some organizational tasks. Katheryn Schneider contributed amazing energy and a thousand ideas. And a third core leader, Sherry Sanders, a well-known and connected community leader, found support whenever we most needed it. During the creation of the Rice Street Gardens, $9,000 of my giveaway money greased the skids for the expensive first-year tasks. There were literally a hundred problems and triumphs during our first year, which swirl in my mind because some are ongoing as I write this. It might be best just to say what we accomplished.

Now, after our second year, we have over 260 gardener families in our 2.5-acre area. We've been told it is the second largest community garden in Minnesota. It's a remarkably international venture with about 90 Karen (from Burma and Thailand), 50 Hmong (originally from Laos), 50 Nepali (from Nepal and Bhutan), and 70 from many other countries as well as multi-generation Americans from the local area. I am slowly communicating with many of them, despite language problems, with at least a smile. Regardless of the problems and concerns, we are creating an exciting, multi-cultural gathering place, a true community with learning and surprising joys every day. We haven't rebuilt Eden, but we've taken one small step for humanity.

Hobbies and Trips

Much of my retirement has been spent on short spasms of learning and experimenting with new hobbies. Some of these have come and gone, but all were my passions at one time. They were not major activities like the Garden or Grandpa Camp, but were interesting moments in my life.

Day trading, or more accurately week trading, was a fun and somewhat profitable pursuit for about one year. Way back in graduate school I competed with a finance student, Boyd Poston, who actually understood companies and the stock market. I professed a system I then called the Harmonic Oscillator Theory and we tracked imaginary investments and returns. My basic idea assumed that investors overreact to events, driving stock prices temporarily too high or too low like a weight on a spring, and eventually returns to equilibrium.

My method seemed to work fairly well, so in 2011 I tried it for real. An investment of about $40,000 turned into $90,000 in a year and a half. I used several rules: (1) choose high beta stocks (ones that move faster-than-average stocks), (2) pick ones that are generally going up, (3) wait for a notable event to drop a stock's price below its 20- and 50-day moving averages, (4) study the stock to be sure nothing had materially changed in the company, (5) buy about $10,000 worth of the shares, and finally (6) wait for them to rebound by 5 to 10 percent. For example, I made tremendous gains after the Japanese tsunami and subsequent reactor meltdown hit the entire market hard. Several of the stocks I was tracking lost over 10 percent, so I bought them. They later gained back all of their original value, plus some. In 2016 I briefly tried this method again, but results were mixed. Perhaps computer-trading systems have made my technique obsolete. I may try stock trading again some day, but I simply have too many hobbies and time is more important to me than money now.

A significant time-eater currently is my love of basketball, specifically our season tickets to the Minnesota Timberwolves. They play 82 games, which uses up about 250 waking hours each year—well worth it, in my opinion. Watching young men soar through the air exhibiting fantastic individual skills, as well as working together as a team, gives me great pleasure. Having played a lot of basketball as a youth, it reminds me of the joy of sports and life. Perhaps I still dream of flying one day. Unfortunately, the Wolves have been consistent losers for about 13 years, which bugs Miriam a lot. But my hope is eternal, and perhaps the excitement of playoff finals, like that provided by Kevin Garnett and company in 2004, will occur again someday before I die.

As mentioned, a huge hobby is photography. It's the first thing I teach the kids in Grandpa Camp. Our photos date back to 1890. The first photo is a scan of Grandfather Hans's baptism certificate, with my translation from Danish attached. Everything has been digitized—my mother's old photos, my father's 9,000 35mm slides, and our entire family's pictures up to 2010. They were all cropped, color corrected, aligned, labeled, rated, faces tagged, and dated. Unfortunately, when Apple changed operating systems about five years ago, the pictures wouldn't automatically upgrade and I was forced to use extraordinary measures with third-party software. All the photos, dates, and titles were saved, but all my previous corrections were lost. So I am still struggling to finalize the corrections, and then I'll still have all the photos after 2010 to sort through. Needless to say, this is an immensely time-consuming process. So working on old photos is officially my hobby of last resort.

I also enjoy creating movies. Starting in the 1990s, the computer applications have improved, and my current favorite, Final Cut Pro X, is just great. It is also a Grandpa Camp regular lesson. Since 1978, at least a dozen new cameras and video cameras have been purchased and used. One major achievement was completing a movie with Miriam's dad, Ken. He had started making 8mm movies while in the Navy during WWII. As a team, we transferred all of his old films from aboard ship, on shore leave in Hawaii and New York, and with his young family to my computer. He sat for hours with me, narrating the movies that I had edited and combined. Those movies are a wonderful reminder of his voice and life.

Over 50 edited and completed movies are now in my archives. They include weddings (e.g., Amy, Kelsey, and her friend Amanda), church programs and concerts, big trips, and numerous family events. Each grandchild is also learning to make special-effect movies with my 20-foot-by-40-foot green-screen cloth.

Recently, aerial movies have been added using my semi-pro drone, including a fly-over of the Rice Street Community Garden, our pond, and

Kelsey and Joe's house and lake. I'm getting requests to do aerial videos for other gardens now. One huge task remains: compiling and editing all my family VCR tapes from 1978 to about 2005. That massive job competes with the photo album work and will probably linger for years.

Learning new languages could be considered a hobby. We first purchased Rosetta Stone around 2005 to help November Paw and her family learn English as quickly as possible. I was impressed by the intuitive teaching method of the program so we subsequently bought the Spanish, Chinese, and Swahili versions, often as a prelude to an upcoming vacation. The Spanish lessons, for me, were a great success because I devoted the necessary time, and of course Miriam was always available as backup. I didn't have as much success with the other languages, but they added background for some wonderful vacations.

We try to take at least one exciting vacation each year. Many are to places south of the border, where some of my Spanish helps (at least to read the signs). We took a Caribbean cruise with Janis in 2006 to Belize, Roatan, and Cozumel. Some beach vacations, but not all, have been with Joe and Kelsey during the dead of winter: Ixtapa in 2007, Puerto Vallarta in 2008, Dominican Republic in 2008, Puerto Rico in 2010, and twice to Cozumel in 2015 and 2017 (Joe likes to ride his bike around the island edge on the great road by the sea).

We also had an extended stay in Central Mexico with Miriam's long-time friends from her time studying in Mexico City, Joan Gmitter and Lynne and Edwin Castillo. I loved our second visit to the beautiful city of Guanajuato ("wanna wattle") where I almost purchased a 9-foot tall metal bird for the entry to our house but Miriam forced me to buy a 30-inch version, now known by all as Quacky.

Five extended trips will never be forgotten. Miriam traveled alone on the first, to Tanzania in 2008. Fearing the medical facilities there, I stayed at home while she traveled with a small group, seeing lions, elephants, and many other animals in open environments. She also visited a small church in Uhominyi village, where she was greeted like a visiting queen. They actually strewed branches and leaves on the path as she approached the small church that Galilee had been supporting as a sister-church. She still wears her Tanzanian outfit occasionally and said that that trip changed her view on life. She was amazed that people who were so poor could be so compassion-ate, hospitable, and joy-filled. This is another opportunity to mention what an amazing wife Miriam is. She spends at least 75 percent of her time help-ing and nurturing other people, me included. I really lucked out.

Our second amazing trip started with a cruise up the Danube River. Starting in Bulgaria, we learned the dispiriting histories of the region. Like Poland, Bulgaria has been one of the walk-over countries in all the great wars, often managed by others. For example, the communists built huge iron factories there, despite having no natural resources to make steel. Those factories and cities are now nearly abandoned. Similarly sad, horrible sto-ries applied to many of the Slavic countries we traveled through: Romania, Serbia, Croatia, Hungary, and Slovakia. Then again, Budapest and Prague were truly beautiful cities. We then flew to Rome to meet Joe and Kelsey for a week in Italy. On their first day we walked the streets of Rome on what Joe called the "March of Death." He had just run Grandma's Marathon in Duluth, preached two sermons, and flown across the Atlantic when we dragged him through the streets and alleyways of Rome for hours. A day or so after his recovery, we visited all the famous sites like the Coliseum and the Vatican. Traveling by train to Florence, a gorgeous city, we learned how to make pasta in a castle while Joe and Kelsey biked through the country-side. Our last stop, Cinque Terre, was a charming group of villages on the Mediterranean, where strolling along the sea cliffs felt timeless and blissful. All in all, it was an excellent vacation.

Our third great trip took us to Australia and New Zealand with the Overseas Adventure Travel (OAT) tour group in 2013. As I write this, I'm wearing a souvenir T-shirt from Uluru (a.k.a. Ayers Rock), the great monolith mountain at the center of the Australian continent. Surrounded by desert and surely a quadrillion flies, this sacred red sandstone mountain is breathtaking, especially at sunset or sunrise. So much about Australia is fascinating—for example, how the desert people live, go to school, and make a living when the nearest neighbor might be 100 miles away. Snorkeling on the Great Barrier Reef and the sites of Sydney were other highlights there.

New Zealand is a country about the area and population of Minnesota, only with rainforests, ocean beaches, glaciers, mountains, and volcanoes. The whole place is intriguing. For example, the city of Rotorua has boiling mud pots, bubbling lakes, and a pervasive sulfur odor. We walked a small valley where colorful long weeds slowly undulated in a crystal clear stream, the most beautiful rivulet I've ever seen. New Zealand is clearly the place to move to if we ever get kicked out of America. In my next life I want to be either a sheep rancher there or to work at the Weta Studios where special-effect movies are made, like the *Lord of the Rings* trilogy.

The fourth excursion that needs to be mentioned is our trip to Japan, China, and Thailand in 2015. We met our tour group in Tokyo, the 3D metropolis I had visited twice before. After seeing the city, I met with former business colleagues, including Ken Kawai, for a fancy dinner, remembering to bring gifts for all, as is the Japanese custom. Everyone seemed happy except the gentleman to whom I gave Minnesota wild rice (he was probably only used to pure Japanese white rice and probably didn't know how to cook it.) The highlight of the Japan part of the trip was driving two-thirds of the way up Mount Fuji with brightly colored fall leaves everywhere.

The Chinese leg of our journey began in Beijing on a foggy, smoggy night. The ultramodern airport there reminded me of the evil spaceship in the first *Alien* movie—very spooky. Everything in China seemed big, as is perhaps appropriate for a country with 1.4 billion people. For example, one doesn't often see single apartment houses. Rather, you observe 30- to 50-story apartments in clumps of 20 to 40 buildings everywhere. Visual highlights around Beijing were the Great Wall, Tiananmen Square, and the summer Olympics grounds with the Bird's Nest stadium. Also, an introduction to the oppressive political rules started to sink in.

Our favorite part of the China trip was around Shanghai, a skyscraper-filled urban area of 24 million people. No site in China better depicts the dynamic economic growth of the country as the riverfront of Shanghai. Fifteen years ago, the east side of the Huangpu River was mostly flat and undeveloped. Today there are dozens of 50-plus-story buildings with lights decorating their exteriors. It's like Christmas in Minneapolis times 100. Advertisements and news can be watched on the lighted sides of the large buildings, especially on a wonderful nighttime river cruise. Most of the city is lit that way.

Of course the crowding and traffic are crazy in Shanghai, which led them to build a bullet train to the nearby cities. Suzhou, 60 miles west, can be reached in 20 minutes. We rode that train, going 200 miles per hour, and watched 50-building apartment groups zip by. The bus ride back to Shanghai took around two hours. The beautiful city of Suzhou (11 million people), on the Grand Canal, is effectively a suburb of Shanghai.

We flew to the Yangtze River for passage through the Three Gorges Dam, a remarkable construction, and proceeded to Chongqing (a city of over 30 million) to see dozens of pandas. A few hundred miles away, Xi'an has the Terracotta Warriors, which were spectacular. Pictures and small exhibitions of these warriors cannot convey the impact of seeing thousands of the soldiers and horses converging in perspective into the distance. Near the warriors in a mercury-filled tomb is the resting place of the first, and probably greatest, Chinese Emperor, Qin Shi Huang. His accomplishments included conquering and unifying the Chinese states, creating a standard system of governance, beginning construction of the Great Wall and Grand Canal, and unifying the language script characters, roads, and monetary system. His tomb is said to have rivers of mercury on a map in the shape of the country, but no one has dared enter because of the mercury vapors.

Other highlights of the journey were the mountain pillars of Guilin, the towering buildings in Hong Kong, and the temples and pagodas of Thailand. It's hard to convey the impressions and education gained on that trip. It is truly a different world over there.

A final memorable tour, so far, was with Peter and Leo to London and environs in 2016. As we were landing, Leo, looking out the window, said it was the best day of his life. For years he had been studying maps of big cities, drawing in airports and new highways where he thought they were needed. A key city he had worked on many times was London. Finally he could see all the roads and buildings in miniature and for real as we came in for the landing.

The first day, the boys forced us to visit LEGOLAND, a large facility near Windsor Castle, with full size LEGO animals, dragons, and miniature city centers from around the world. Throughout the trip we told the boys where we were going and they would plan the Tube route. In general, all types of transportation there captivated them. Other highlights included the Tower of London, the HMS *Belfast* tour, the London Eye, Trafalgar Square, the War Museum and Transportation Museum, day train trips to

Cambridge, Greenwich, and Hampton Court. Most surprising was their rapt attention to the TV and newspaper accounts of the Brexit voting, which occurred during our week there. Miriam and I did not expect that they would often grab a free paper to read during each Tube ride. The trip was a hoot. We look forward to going with Elaina, Heidi, Linus, and Tim on similar trips in our future.

Electric

From the title you might infer this section is about my "electric" personality, but not so. It's about the sequence of events that recently led to a $2 total household energy bill for us during the month of May. The trigger to all this was the advent of the Tesla, an all-electric car, about four years ago. Son-in-law Joe and I had been tracking Tesla's progress for a few months, and when they finally opened a Minneapolis sales and service center in May 2013, I decided to take a test ride. We were attracted by both the no-gasoline idea and because the car was very fast, slightly faster than my Porsche 911. The test ride was amazing. The car was big and had massive cargo space (two trunks in the back and one in the front, not to mention a missing transmission hump in the middle). The engine was smaller than a golf bag and the batteries were spread across the car bottom, providing stability in turns. The car had a 265-mile range before needing a recharge. In June 2013 they won the Car of the Year award from several magazines and the National Highway Traffic Safety Administration (NHTSA) gave the car a 5.0 perfect safety rating. Also, the car was cool looking and had a giant, 17-inch touch screen control panel with automatic software updates. A month later I ordered the car, and a month after that Tesla had built the car and shipped it to Minneapolis. The price was more than I had ever spent on a car, with some relief at tax time, but it was well worth the expense. We plan to keep the car for about 10 years and then buy a new one with self-driving capability as I enter my dotage years. I only wish I had bought stock in Tesla sooner. The stock price rose from $20 in early 2013 to about $350 today. It is very clear to me, as an engineer and futurist, that the advantages of electric cars are so great that they will eventually replace the internal combustion engine.

Driving to Champaign, Illinois, for a memorial meeting when my friend and graduate advisor, Andy Anderson, died became the first test for finding charge stations on the open road. It turned out not to be a problem because the car pointed us exactly to the four needed stops along the way. Under normal use, it is a great joy to have a "full" car every morning and never have to stop at a gas station (except to get gas to fill the lawn mower or Miriam's Toyota Prius). The only remaining roadblock for Tesla and others is the cost and availability of Li-ion batteries, which is rapidly improving.

Although not my prime consideration, the low carbon footprint of the Tesla was an important factor, which ultimately led us to our second great ecological move, buying solar collectors for our house. Solar energy research was one of my first jobs at Honeywell back in 1973. The goal of those early solar panels was to gather energy to heat homes and buildings. But today the technology for solar-electric conversion panels has progressed so far that they are economically competitive with electric companies and are a good investment. We are fortunate to have a large, south-facing roof on our home and few trees to block the sun. A sales engineer approached us from All Energy Solar company to walk us through the benefits. In short, we spent $40,000 for our system and received $13,000 back on our taxes for a net cost of about $27,000. We have 36 solar panels, which provide about 70 percent of our energy over the course of a year (including the energy needed to power the Tesla electric car). The collectors are generating about $1,700 worth of electricity per year. With an additional $1,000 per year incentive from Xcel Energy, we are getting a 10 percent return on our investment.

We are very happy with both the economics of our decisions and the positive (albeit small) contribution to controlling global warming. As a bonus, our city of Shoreview just presented us with a Community Green Award as a unanimous choice for 2017, and we proudly display our plaque by the front sidewalk.

Miriam

One constant through much of my life's journey has been Miriam. How do I love her? Let me count the ways: as a friend and lover, mother to our children, helper in dangerous times and routine life, devoted foundation, in-spiration, snuggler, and mate. That's nine ways, at least. We are almost the perfect complementary pair, she worrying about the details and other people and my mind always in the clouds and future. We are both independent and united. The system might crack if either of us leaves, but for now we are very happy. It's a joy to grow older with her, not the getting older part but feeling whole and bound as we go.

We've had scares along the way, during childbirth and when she got breast cancer and struggled through surgery and chemo. She also recently had a hip repaired, which caused some concerns. She was always close as I faced my mortality, like bleeding internally, kidney removal, spleen shrinking, and less serious medical procedures such as carpal tunnel sur-gery. We mourned together when our parents died: Ed in 1982, Fern in 2005, Helen in 2010, and Ken in 2015. Both of us are so matter-of-fact that our lows are never debilitating and our highs are slightly muted. We don't say "I love you" often enough, but the certainty is always there. I won't say she is the best spouse in the whole world (as her parents often said), but she is within striking distance. Through it all, I love her and I always will.

So What Should I Do Now?

If my prediction comes true and I croak at age 81, it means there are about nine years left to finish all my projects and anything new that comes up. What to do? Of course, living past 81 is certainly possible and desirable, but who knows? Let's consider the problem.

The first job, I suppose, is to finish my existing projects. Here's the list:

- Archive all the family photos, over 80,000 and rising
- Edit and finalize the family movies
- Get all the information to family members
- Finish and publish this book and the science fiction novel written in the early 2000s
- Gradually transition the Rice Street Garden to the gardeners them-selves while teaching them management skills

- Continue and improve Grandpa Camp teaching and love for all the grandchildren
- Clean up and simplify our house

Finishing these projects will necessarily lead to additional tasks. For example, publishing the books will require marketing, building a good website/blog, and possibly starting a sequel. Editing movies could lead to doing more professional aerial drone movies. I think expanding the Grandpa/Nerd Camp to paying, non-family people might be fun. Writing a book about Grandpa Camps is another intriguing possibility. Helping others start new community gardens would be a good service activity.

So what makes a worthwhile life? There should be an equation for this, perhaps something like: [improve the lives of people] x [the number of people impacted] x [how long the benefit lasts]. Some people stop trying to have an impact or learn new things. A few in high school assume they know enough already. But I will never stop learning, doing valuable activities for others, and trying to make the world a little better.

Many people simply try to be happy, especially in retirement. Others focus on the afterlife, trying to be right with God. I simply don't know how I can be happy without having meaningful ways to help others. And I don't know how to be right with God by only praying and reading the Bible. My faith tells me that Jesus already died to reconcile humanity to God, so I need to do more. I want to learn and do materially positive things until I'm no longer able.

In particular, I would like to have one more great adventure, something that makes people smile and remember. As aging kicks in, it becomes increasingly difficult to do something big when your joints ache. High jumping is out. Perhaps there is something inspirational I can still do. But what?

We could take more trips: an around-the-world cruise, a sail to Antarctica, a visit to Yellowstone or Venice before it sinks, etc. We could take the entire family on a big trip, a learning or research adventure for the grandkids. I could run for political office, just for the fun of it. We could get a pet dog (something we've never done). I could paint or photograph stunning images, bringing more beauty into the world. I could help my grandchildren do important science projects so they can get into any university they choose and have lives of significance. We could give away our money (as we are already doing with Augsburg, Caltech, family, and Galilee) to something bigger.

The main thing is to know the general direction of the future. That's how I ran the research labs at Honeywell. And when opportunities emerge along that path, to quickly seize them. Check in with me in ten years to see how I've done. Wish me luck.

Flying

So how does it all add up—was I a success at life? On the one hand, I never became famous like my heroes: Galileo, Einstein, Dick Fosbury, Elon Musk, or national leaders or movie stars. However, I've accomplished more than most, and plenty considering my beginnings. I'm satisfied.

In one sense I've conquered my introversion. The ability to mingle at parties and speak before large crowds slowly developed over the years. I have to gird myself, put on a new face, and soldier through, but I can do it. Afterward, I feel tired and need to recover. But I can act like an extrovert for short periods. My humor saves me. I make it a game. In fact, all of life seems like a game at times and I will never grow old in my mind.

I had a lot of help: loving, supportive parents, teachers who saw me as a friend, people who instinctively trusted me and carried out my plans. In the era I grew up, being a white male undoubtedly helped. Possibly my being shy and quiet was valuable. I never seemed a threat or serious competitor to others, but they knew I was genuine and competent. Why was I made school police captain in grade school or head of the National Honor Society in high school? I didn't vie for such opportunities. They just seemed to happen.

I think my introversion was a major advantage. In the morning before getting up, while taking walks or cutting the grass, in the quiet moments of life, I thought about the problems and opportunities around me. I analyzed my bosses, learned from training classes, solved problems, and plotted the future, often without really trying. Being thoughtful and alone at times was useful to me and likely a great advantage in life. People wonder how I avoided disasters and succeeded in most areas, and I think it came in the quiet times, when my mind could dream and analyze.

So if you are an introvert, be of good cheer; it may be one secret to a fun life.

Did I learn to fly, like in my childhood dream? So much of life is routine: getting dressed, cooking and eating, exercising, earning a living, learning, maintaining things, and sleeping. There's not a lot of time for changing the world. There were moments of greatness in my life: the state track meet; graduating from Caltech; being one of the world leaders in low temperature physics, solar energy, and satellite protection; getting promoted seven times; leading a $200 million organization with 2,000 employees; helping win an $11 billion contract; marrying a caring, gifted wife and fathering wonderful daughters; loving six joyous grandkids; and creating a garden for 260 families. Perhaps I *was* flying. Perhaps it's all a dream, and one day I'll wake up and God will say I should have done more, that I was one of His blessed ones and could have done so much more for others. You do what you can. I'll keep at it.

Ron leaves Honeywell in January 2000 with assisstant Carol Warne.

Machu Picchu, August 2000

Kelsey graduates from Mounds View High School, 2001

Amy graduates from Caltech with a B.S.
in Engineering and Applied Science, 2001

Amy's wedding to Dan Fisher,
August 4, 2001

Thanksgiving at Helen and Ken's, 2003

Ron is ready to photo on the zip lines of Costa Rica.

Jell-O disaster.

Ron leads a town hall meeting on education.

First grandchild, Peter, is born March 8, 2004. Pictured with Dan and Amy.

Great Grandma Fern loves Peter.

Peter loves Great Grandad Ken.

Early reader.

Punting on the Cam River with Kelsey during her semester abroad.

Kelsey graduates from Valpariso University, May 2005

Grandson Leo is born, March 12, 2006

Building Habitat houses in Mexico, 2006. Pictured in Guanajuato.

Kelsey is engaged to Joe Midthun.

Joe and Kelsey are married, October 6, 2007

Peter is an angel.

Leo and Ken both love maps.

November and December Paw become citizens, 2013

Miriam meets students we support in Tanzania, 2008

Janis marries Fred Behrens, 2009

Joe is ordained.

Grandson Tim is born, July 24, 2010

Tim, the diligent chess student, 2017

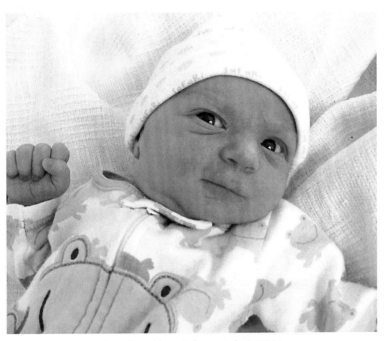

Grandson Linus is born, July 3, 2014

Linus, the future fireman, 2017

All the grandchildren, 2017: Linus, Leo, Peter, Elaina, Tim, and Heidi.

Peter painting, Grandpa Camp 2013

Leo's special effects movie, Grandpa Camp 2013
"The Time Machine"

Leo and Peter build a six-inch telescope,
Grandpa Camp 2014

Peter dissects a cow eyeball, Grandpa Camp 2014

Tim's fetal pig anatomy lesson, Grandpa Camp 2017

Peter studying Brexit on the Tube, Grandpa Camp 2016

Elaina is born, May 28, 2013

Elaina at Grandpa Camp, 2016 and 2017

Heidi is born, May 24, 2016

Australia and New Zealand trip, 2015. On the Great Barrier Reef.

Japan, China, Thailand trip, 2016

Electric car and solar roof.

Rice Street Community Garden.

About two-thirds of the garden shown in picture.

Retired but working on multiple fascinating projects, Ron would still rather play with his grandchildren than argue with overly opinionated adults. He lives in Minnesota with his patient and tolerant wife, Miriam, who tries to keep his feet on the ground while his head remains in the stars. Find him online at ronsreadingroom.com.

The text of *An Introvert Learns to Fly* is set in Adobe Caslon Pro. Book design by Sarah Miner. Composition by Bookmobile Design and Digital Publisher Services, Minneapolis, Minnesota.